Brenna

What Makes a Computer

I0009761

Brenna C. O'Brien

What Makes a Computer Genius?

Developing Computer Technology Talent

VDM Verlag Dr. Müller

Imprint

Bibliographic information by the German National Library: The German National Library lists this publication at the German National Bibliography; detailed bibliographic information is available on the Internet at http://dnb.d-nb.de.

Cover image: www.purestockx.com

Publisher:
VDM Verlag Dr. Müller Aktiengesellschaft & Co. KG
Dudweiler Landstr. 125 a, 66123 Saarbrücken, Germany
Phone +49 681 9100-698, Fax +49 681 9100-988, Email: info@vdm-verlag.de

Produced in USA and UK by:
Lightning Source Inc., La Vergne, Tennessee, USA
Lightning Source UK Ltd., Milton Keynes, UK

ISBN: 978-3-8364-3641-0

ABSTRACT

This study explored the development of computer technology talent (CTT) in the field of gifted education. Based on Feldman's (1994) co-incidence theory, the four time frames have crystallized at the end of the 21st century to produce conditions for CTT to emerge: (1) individual life span, (2) development of the field, (3) historical and cultural trends, and (4) evolutionary time. Within this framework the research question was asked: "What cognitive and affective qualities and life events mark the development of CTT?" Using qualitative study methods, interview data were collected from four different age groups and time periods: Historical (Sample 1), Snapshot (Sample 2), Longitudinal (Sample 3), and Contemporary (Sample 4). Gagné's (2003) Differentiated Model of Giftedness and Talent (DMGT) was used after the initial analysis to structure the data into the categories of natural abilities, intrapersonal catalysts, environmental catalysts, and talent activities. Findings were discussed in terms of developmental patterns and recurring trends between samples, reflected in their histories with computing, environmental support, and key educational experiences. Results provide evidence that CTT should be recognized as a distinct talent area within the gifted field. Suggestions for educational policy and practice are made for teachers and parents of children with CTT. Alternative assessments and resources need to be made available in schools to recognize this new way of thinking in the digital age.

ACKNOWLEDGEMENTS

I would like to thank all of the people without whom this book would not have been possible. To Dr. Kevin Besnoy, thank you for your encouragement and support, and for giving a doctorate student a chance to make a real difference in gifted education. Thanks to Dr. John Monberg for challenging me to explore the deeper issues related to technology and society, and for always pushing me to go further with my research. The staff of the University of Kansas library deserves a special mention for always taking the time to fulfill my endless requests for obscure technology articles from around the world.

To my committee members: Thank you Dr. Karen Jorgensen for your insight and experience, and Dr. Phil McKnight for your kindness and willingness to see my project through from proposal to completion. Dr. Nancy Baym, thank you for your inspiration and for teaching me what it takes for a person to have a true dedication to qualitative research. I knew I could always count on you, Dr. Marc Mahlios, and I thank you for your advice and encouragement over the years that helped guide me through those scary halls of academia. And finally, to my advisor Dr. Reva Friedman-Nimz, your genuine concern for me has been the shining beacon that has motivated me throughout my academic career. Thank you all for sharing your wisdom and knowledge with me as I completed my last step of my doctorate degree.

This book is dedicated to my very patient husband, Blake Washer, who has been understanding and supportive throughout this long and winding journey, and to my beloved father, Patrick O'Brien, who taught me everything I know.

TABLE OF CONTENTS

                                                                        Page

Abstract   2

Acknowledgements   3

Table of Contents   4

List of Tables   9

List of Appendices   10

CHAPTER 1

    Introduction   11

    Problem   12

    Purpose   13

    Definition of Terms   14

    Disclosure Statement   15

    Summary   17

CHAPTER 2

    Introduction   18

    Theoretical Framework   18

    Feldman's Co-incidence Theory   20

    Individual Life Span   21

        Gagné's Developmental Theory   21

    Developmental History of the Field   32

        Universal-to-Unique Continuum   33

    Historical and Cultural Trends   37

    Evolutionary Time   42

    Summary   44

CHAPTER 3

Introduction 45

Research Objectives 45

Research Question 45

Qualitative Design 46

Rationale for Samples 47

Sample 1: Historical 48

   Setting 48

   Participants 49

Sample 2: Snapshot 50

   Setting 50

   Participants 51

Sample 3: Longitudinal 52

   Setting 52

   Participants 53

Sample 4: Contemporary 54

   Setting 54

   Participants 55

Data Collection 56

Validity and Reliability 58

Analysis Procedure 59

Ethics 61

Summary 61

CHAPTER 4

Introduction 62

Sample 1: Historical 63

   Natural Abilities 65

     Intellectual 65

     Creative 68

Socioaffective 71

Intrapersonal Catalysts 74

    Motivation 74

    Volition 76

    Self-Management 77

    Personality 79

Environmental Catalysts 81

    Family 81

    School 82

    Peers 84

    Media 86

Talent 87

Predictions for the Future 88

Summary 90

Sample 2: Snapshot 92

Natural Abilities 93

    Intellectual 93

    Creative 95

    Socioaffective 96

Intrapersonal Catalysts 99

    Motivation 99

    Volition 100

    Self-Management 102

    Personality 104

Environmental Catalysts 105

    Family 105

    School 107

    Peers 109

    Media 111

Talent 112

Predictions for the Future 115

Summary 118

Sample 3: Longitudinal 120

    Natural Abilities 121

        Intellectual 121

        Creative 123

        Socioaffective 124

    Intrapersonal Catalysts 126

        Motivation 126

        Volition 128

        Self-Management 129

        Personality 131

    Environmental Catalysts 132

        Family 132

        School 134

        Peers 136

    Talent 138

    Predictions for the Future 139

    Summary 141

Sample 4: Contemporary 143

    Natural Abilities , ... 143

        Intellectual 143

        Creative 146

        Socioaffective 148

    Intrapersonal Catalysts 150

        Motivation 150

        Volition 153

        Self-Management 154

        Personality 156

Environmental Catalysts 158

    Family 158

    School 160

Peers 163

Talent 165

Predictions for the Future 168

Summary 169

CHAPTER 5

Introduction 172

Conclusions 172

    Natural Abilities 173

    Intrapersonal Catalysts 174

    Environmental Catalysts 178

    Talent Activities 181

    Co-incidence Theory 183

    Universal-to-Unique Continuum 185

Limitations 188

Implications 191

    Policy 191

    Practice 194

Future Research 197

Summary 199

References 200

Appendices 222

List of Tables

Table 1. Information Table for Sample 1: Historical

Table 2. Information Table for Sample 2: Snapshot

Table 3. Information Table for Sample 3: Longitudinal

Table 4. Information Table for Sample 4: Contemporary

Table 5. Trends for Natural Abilities

Table 6. Trends for Intrapersonal Catalysts

Table 7. Trends for Environmental Catalysts

Table 8. Trends for Talent Activities

List of Appendices

Appendix A. Participant Interview Questions

Appendix B. Thumbnail Sketches Developed from Pilot Study

Appendix C. Pilot Study Interview Questions

Appendix D. Thinking Style Rating Scale

Appendix E. Pilot Study Follow-Up Questions

Appendix F. Parent Interview Questions

Appendix G. Teacher Interview Questions

Appendix H. Focus Group Interview Questions

Appendix I. Computer Technology Talent Study Information Sheet

Appendix J. Informed Consent Statement

CHAPTER 1

In the past couple of decades, early adopters of technology have usually been children and teenagers, leaving many parents and teachers in awe of their knowledge and skills (Tapscott, 1999). The entire culture has been affected by these changes, as noted by Cross (2005): "This generation of children who are now in their teens has become so technologically savvy that being passionate about technology is becoming more commonplace" (p. 26). According to recent PEW data, 87% of children 12-17 are on-line, and the percentage continues to increase (Fox & Madden, 2006). It seems that now more than ever it is important for schools to recognize young people demonstrating exceptional computer technology savvy, and provide pertinent talent development opportunities if these individuals are to realize their full potential.

With the integration of computers and other technology into almost every U.S. classroom, different levels of student ability need to be recognized and specific accommodations made for students with extraordinary ability. Without a challenging environment, it has been noted that many intellectually gifted students can become bored and lose interest in school achievement (Winner, 1996). It is logical to assume that the same would apply to young people whose talents center on computer technology. A qualitative research design was employed to explore natural abilities, dominant intrapersonal qualities (e.g. internal motivators, personality attributes), and external catalysts. Nodal developmental points were identified to create a portrait of critical technology experiences. Interviews conducted with four samples from different age groups and time periods were used to compare and contrast the emergent developmental patterns.

11

Problem

Computers have become an integral part of society, but there are concerns about whether the next generation of children is being effectively taught the technology skills and knowledge necessary to compete on a global scale. An article by Frauenheim (2005) was recently published titled, *Can Johnny Still Program?* In this report, David Patterson, a computer science professor and president of the Association for Computing Machinery (ACM), talked about the need for encouragement of computer technology talent in America:

> After U.S. students earlier this month made their worst showing in the 29-year history of the ACM International Collegiate Programming Contest, Patterson and others are wondering whether the United States does enough to encourage programming talent. (p. 1)

According to him, not enough emphasis has been placed on encouraging programming for young gifted students with an inclination towards technology. Another issue important to this study is how rapidly the thinking patterns of children are being changed because of technology. In the educational magazine *Edutopia,* McHugh (2005) referred to the current generation of students as 'iKids,' and wrote that, "Today's brains are shaped by various information streams – sometimes referred to as memes – constantly popping and sparking and competing for attention. This new generation of digital learners take in the world via the filter of computing devices" (p. 33). Studies on how this learning is related to intellectually gifted children and computer technology talent are just beginning to emerge (Heinzen, in press).

Some research has been conducted on how children develop technology skills (Healy, 1998; Tyler, 1998; Cohen, 2001; Gillespie & Beisser, 2001; Chiero, Sherry, Bohlin, & Harris, 2003) but very few studies have been conducted that consider technology skills through the unique lens of gifted education. The personal qualities and environmental settings necessary to produce talented individuals have also been examined extensively in the field of giftedness, but there is a paucity of research that specifically examines the developmental stages of technology talent. This study was designed to permit these gifted individuals to explain and explore their development in the domain of computer technology in their own words.

Purpose

This study's primary purpose was to examine the personal characteristics and life experiences that affect the development of students with computer technology talent. Interview snapshots were captured of different samples to trace the development path of this talent area. The more information known about the intellectual and personal qualities of this group of people, the easier it will be to facilitate their learning in the educational system and advise parents on ways to encourage this development.

How does simple computer interest evolve into computer talent? Bozionelos (2001) offered this definition of *computer interest*: "Computer interest encompasses positive emotions and cognitions regarding the prospects of acquiring knowledge about computers and actually engaging in interaction with computers" (p. 429). Talented individuals build on those positive feelings and continue to explore the computer system and create new ideas that go beyond the average person's scope. These exceptional individuals can hone their skills in the area of computers and technology and have the potential to use their talent to be successful. The factors that influence this development are the subject of this study.

Because there are few standards currently in place for computer technology talent, an emergent and open-ended approach was taken. The research objectives of the proposed study were: (1) to describe in rich detail the distinguishing intellectual and personal qualities of adults and adolescents demonstrating computer technology talent; and (2) to trace the developmental path of computer technology talent in the lives of adults and adolescents. By examining interviews from four samples, a multi-faceted description of this talent area was compiled to add to the body of knowledge in the world of gifted education and technology studies.

Definition of Terms

Development – In this study, Feldman's (2000) definition of development is used within the framework of gifted education: "Development includes all of the ways in which general and specific potentials may be brought to expression through conditions and contexts evolved and utilized by cultures" (p. 13).

Computer Technology Talent (CTT) – This term is used to refer to the phenomenon being studied and denotes high knowledge of and ability with computer technology, above and beyond same age peer groups. Advanced computing skills at a young age comparative to other children is one example of CTT.

Technology – To maintain a manageable focus for this study, the term technology refers specifically to computer systems, both hardware and software. Information and Communications Technology (ICT) is also included here as a facet of computer use. Other technological devices (video games, digital recorders, cell phones, etc.) are discussed specifically when they are brought up in different contexts.

Talent – The definition of talent utilized in this research was posited by Gagné (2003):

"Talent designates the superior mastery of systematically developed abilities (or skills) and knowledge in at least one field of human activity to a degree that places an individual at least among the top 10% of age peers who are or have been active in that field or fields" (p.60). Emphasis is placed on comparing skills of an individual to the skills of same age peer groups.

Gifted –The current federal definition of gifted is located in the *Elementary and Secondary Education Act* (2001): "Students, children, or youth who give evidence of high achievement capability in areas such as intellectual, creative, artistic, or leadership capacity, or in specific academic fields, and who need services and activities not ordinarily provided by the school in order to fully develop those capabilities (Title IX, Part A, Section 9101(22))" (p. 544). The current Kansas state definition is based on legislation by the *Kansas State Department of Education* (2004): "Performing or demonstrating the potential for performing at significantly higher levels of accomplishment in one or more academic fields due to intellectual ability, when compared to others of similar age, experience, and environment (Chapter 91, Article 40(1))" (p. 1).

Disclosure Statement

When focusing on a solely qualitative study, the researcher's experiences and biases are inescapable. Instead of trying to eradicate this subjective perspective it should be embraced, for it is at the very heart of qualitative research to filter information through the interviewer's own life. As Coleman (2001) states, "The person doing the study is the prime research instrument in interpretive scholarship. All data are comprehended through

15

the researcher's mind as he or she constructs the meaning of the participants from the data" (p. 171). My interest in the subject of technology talent was brought about mainly by my own experiences growing up with computers, and my fascination with all things related to the 'magical grey box.' I had contact with different aspects of gifted education during my school career, but I chose to avoid any formal differentiation and remained integrated with the rest of the student body. For the most part I was left on my own to explore the subject areas that interested me. I am curious how my education might have been different if more gifted resources had been made available for me through the school system.

I remember fondly sitting on my father's lap in the mid-1980s staring at the black screen and green text of our brand new Apple IIe. My romance with computers started then, and I enjoyed the challenge of text games and puzzles. I joined the after-school computer club in fourth grade, but school was not the place where I learned the most about computing. I made it a priority for my parents to continue to upgrade our computers so that I could play the newest games and find out about the latest programs. During my freshman year of high school, we got a 56k modem and I was able to connect with local Bulletin Board System (BBS) servers and chat with fellow computer enthusiasts. The internet was so exciting and it felt like my own special place because so few of the other students at school knew about it at that point. I spent hours building websites with Hypertext Markup Language (HTML) and connecting with people all over the world who were just as eager to talk about computers as I was.

After completing my undergraduate degree in Psychology and Secondary Education, I sought employment in a public high school in an urban setting. I faced many frustrations trying to integrate technology into my classroom, mainly because of inadequate resources and rigid school policies. I found that there was a large difference between my own experience and the skills and knowledge of the older teachers with whom I worked. If I was going to make a larger impact on how schools use technology, I knew I would have to advance my own education. While completing my graduate studies, I had the opportunity to teach a class in a computer lab titled Instructional Technology in Elementary/Middle Education. It was my goal to educate the next generation of classroom teachers about the benefits of technology, so they could understand how their students learned differently while using computers. While teaching this class, it became apparent to me that these future teachers did not know how to distinguish between different types of technology learners. My experience with this class informed my decision to research this area of education more closely.

Summary

Chapter 1 presents the problem and purpose of this study, and reveals the background of the researcher in relation to the current research topic. Chapter 2 takes a more in-depth look into the theories of David Feldman and Françoys Gagné, and examines the literature surrounding computers, education, and society. Chapter 3 explains the qualitative methodology used and describes the different settings and participants analyzed for this study. Chapter 4 examines the interviews for each of the four samples from different time periods and age groups. Chapter 5 presents the final analysis and discussion of the results.

CHAPTER 2

Literature Review

The body of research directly related to gifted education and computer technology talent is very slim. Therefore, to support the purpose of this study the literature review was structured into the four areas of Feldman's (1994) co-incidence theory. This theoretical framework was chosen because it presents a powerful de facto argument for the emergence of and development of computer technology talent at this particular point in human history. Previous research focusing on the juncture of children, technology, and education was examined to establish a foundation for this new area of talent development to be added to the body of already recognized talent domains.

Theoretical Framework

A variety of foundational theories have evolved in the field of gifted studies, causing disagreements between scholars on which is the most valuable path for research. Some theories have emphasized a generalized conception of giftedness, where aptitude tests are involved to produce clearly demarcated lines separating gifted from non-gifted individuals. For example, Assouline (2003) provided a timeline for the cognitive and educational assessment of children that explained different standardized tests used over the years to determine intelligence. She showed how many of the tests like the Stanford-Binet Intelligence Scale and Wechsler Intelligence Scale for Children (WISC) were originally based on Spearman's general factor (*g*) of intelligence that he developed in 1904 (p. 127). With these tests, children's intelligence quotient (IQ) could be determined, and if they scored a 130 or above, two standard deviations above the norm, then they were classified as *gifted* (Gottfredson, 2003, p. 28). However, many educators have recognized that only looking at a general concept of intelligence could end up unwittingly overlooking students that have specific talents.

Alternative theories have been developed to expand on the previously established definition of intelligence. Gardner (1983) developed a theory of multiple intelligences where he named seven distinct ways of knowing and learning (linguistic, logical-mathematical, spatial, bodily-kinesthetic, musical, interpersonal, and intrapersonal). However, Gardner's theory was mostly implicit and there has been little research in the field to support his conclusions. In 1999 Gardner revised his multiple intelligence theory, but he still did not recognize technology as an important area. Specifically related to giftedness, Tannenbaum's (2003) model recognized the importance of external factors on a person's achievement level, but ultimately it did not address the developmental issues related to this study. Renzulli (2003) proposed an alternative three-ring definition of giftedness containing above average ability, creativity, and task commitment. He placed this model on top of a pattern of internal and external influences that he labeled "Operation Houndstooth" (p. 79). The aspects he described overlap with many concepts present in this study, but again this theory is not developmental in nature.

The area of research explored in this study relates to the development of computer technology talent. Within gifted education, there are different definitions used for the terms *gift* and *talent*. Van Tassel-Baska (2001) tried to draw a distinction between these two related concepts in her own words: "Giftedness is characterized as distinctly above average functioning in domains of ability (aptitude), and talent is characterized as distinctly above average achievement in fields of performance" (p. 20). As a rallying cry to the gifted education field, Feldhusen (1996) encouraged researchers to, "Move from the global conception of giftedness to the growing emphasis on identifying and developing talent in students" (p. 66). Another researcher who supported the notion of talent development was Gagné (2003), and his theory about how gifts grow into talents was used to structure the data in this study. He was also chosen because he was one of the first theorists to recognize technology as a specialty field and acknowledge its merit as a legitimate talent area. Before individual development was investigated, the overarching historical and cultural developments were examined to see why this area of interest was emerging at this point in time.

Feldman's Co-incidence Theory

Conditions in the last few decades have come about in the field of computer technology that have laid the groundwork for talented individuals to emerge. The development of this specific talent fit Feldman and Goldsmith's (1986) original co-incidence theory of talent emergence. They explained that the term *co-incidence* represented, "The melding of the many sets of forces that interact in the development and expression of human potential" (p. 11). In one early metaphor, Goldsmith (1990) likened these forces to the layers of an onion, comprised of:

> The individual child, individual enablers, and institutional enablers must be coordinated and then interact productively over time if prodigious talent is to develop. If one of these factors is missing, or even mistimes, it becomes far less likely that a prodigy's talent will continue to flourish. (p. 23)

In a later article, Morelock and Feldman (2003) identified the four factors whose juxtaposition created a situation ripe for prodigious talent to emerge in a particular domain: "(1) The individual's life span, (2) the developmental history of the field or domain, (3) historical and cultural trends impacting individuals and fields, and (4) evolutionary time" (p. 462). It appears that a convergence of all of these elements has

20

occurred at the end of the twentieth century. Feldman's co-incidence theory opens a conceptual door to the possibility of computer technology being legitimately recognized as a separate talent domain. Each of these four factors is explored in depth in relation to computers and technology.

Individual Life Span.

According to Morelock and Feldman (2003), the lifespan of the individual included (a) biological advantages, (b) the point in a child's development when he or she is introduced to a domain, and (c) the amount of parental support and outside resources that were available at a young age (p. 462). In an earlier article, Feldman and Benjamin (1986) stated that the strengths of a person were, "Due to the continuous interaction of an individual with various potentials and a world with various possibilities. If these processes of interaction lead to high level performance, then it is appropriate to speak of giftedness" (p. 287). They emphasized that only a small percentage of individuals were able to fully realize the extent of their creative and intellectual abilities given the right circumstances.

Gagné's Developmental Theory.

To further explore the lives of individuals who have demonstrated computer technology talent, the research of Gagné (2003) was examined to trace the development of a person's gifts into talents. This developmental process included an individual's natural abilities (intellectual, creative, and socioaffective) and intrapersonal characteristics (motivation, volition, self-management, and personality), which are affected by the external influences of family and other environmental factors. The specific categories within this theory are explored further in Chapter 4.

21

Under the heading of natural abilities, the first category describes intellectual abilities such as logical reasoning, memory, verbal and spatial skills, and metacognition. Gagné (2003) wrote: "These natural abilities manifest themselves in all children to a variable degree. It is only when the level of expression becomes outstanding that the label *gifted* may be used" (p. 62). Some studies have been conducted that attempt to discern what intellectual skills are necessary for working with computers and technology. In a work setting, MacPherson (1998) administered professional technicians a series of tests before presenting them with a practical problem to troubleshoot and solve. His results showed that the cognitive skills score was one of the strongest predictor of technological problem solving skills, second only to years of experience (p. 18). The results were even stronger when the factors of intellectual abilities, critical thinking skills, and years of experience were combined. The capacity to problem solve appeared to be a crucial intellectual ability necessary to successfully work with technology. An individual's level of understanding of technology can also be described on a continuum with varying degrees of involvement. Custer (1995) defined different types of technological thinking to show the multiple dimensions present in the current society:

> It is apparent then that there exists a continuum of knowledge that draws from practical experience with designing, developing, troubleshooting and repairing technological artefacts. At one end of the continuum is the highly systematized and formalized knowledge of the engineering profession. At the other extreme is the tacit knowledge of skilled tradespersons and artisans. (p. 230)

Some technological procedures have simpler algorithmic solutions while others require a more heuristic or experimental approach. It is clear that individual abilities and experience affect how one learns technology, as well as what level of expertise they can hope to achieve. Higher-level thinking should continue to be promoted for unique

breakthroughs to continue to occur in the field of computer science.

Another important natural ability important to this study is creative thinking.

There is a long history of research in creativity intertwined with the study of gifted and

talented individuals (Torrance, 1962; Treffinger, Renzulli, & Feldhusen, 1975;

Hennessey & Amabile, 1988; Taylor, 1988; Walberg, 1988; Torrance; 2000). In specific

relation to technology, Saunders and Thagard (2005) examined how creativity was

displayed in the field of computer science. They analyzed interviews with practicing

computer scientists to find out what type of creative thinking was used by them while

programming and problem solving. The respondents talked about switching into an

intense mode of creativity that involved scribbling, doodling, brainstorming,

experimenting, and having animated conversations with their peers about problems while

trying to solve technology problems (p. 160). A creativity scholar, Csikszentmihalyi

(1996) described the feelings produced by creative construction:

> The reason creativity is so fascinating is that when we are involved in it, we feel that
> we are living more fully than during the rest of life. The excitement of the artist at the
> easel or the scientist in the lab comes close to the ideal fulfillment we all hope to get
> from life, and so rarely do. (p. 2)

It is these types of creative forces that were present in these expert programmers in

Saunders and Thagard's (2005) study. In Graham's (2004) book, *Hackers and Painters*,

the aesthetic value of computer code was discussed as well. These hackers described

ascending to a new level of purity and beauty while coding, which Graham likened to the

artistic feelings of great painters. Efficient and clean code could provide the same

satisfaction to them as hearing a beautiful piece of music or looking at a masterpiece

painting: "Great software requires a fanatical devotion to beauty. If you look inside good

23

software, you find the parts that no one is ever supposed to see are beautiful too" (p. 29).

Creative thinking, imagination, and having an artistic sensibility are important factors in

the development of computer technology talent.

In the area of socioaffective abilities, the stereotypes of *geek* and *virgin* were

characteristics often considered to appear in conjunction with extreme intelligence

(Coleman & Cross, 1988; Hardin, 1989; Toblin, 2001). Historically, the characteristics

used to define someone as a *nerd* could cause stigmatization and isolation. In his

longitudinal study of middle school students, Kinney (1993) stated: "Some nerds were

singled out for their superior academic performance. Others were viewed primarily as

having low levels of social skills (e.g. being shy, nervous, or embarrassed around others)

and dressing out of fashion" (p. 27). In relation to social abilities, Holland and Astin

(1962) determined that often those individuals most suited for highly creative work were

usually undersocialized. Since creative problem solving is so important to the field of

computer science, it seems like some individuals with CTT had to sacrifice social

interactions to develop their talents. This current study tried to explore beyond the

stereotypical image of a computer geek to discern what other socioaffective qualities

distinguish this group. Cross (2005) noticed how the stereotype of a computer enthusiast

had evolved over the last few decades:

> The children who would have been typically identified early by their peers as a nerd
> or geek, and experienced stigma of giftedness and limited social acceptance, are now
> less likely to experience being an outcast. This is due to the growing awareness that
> being a nerd or geek could actually have a positive outcome.
> (p. 27)

After the technology boom of the 1990s, having computer knowledge and skills became

more popular, and it was more socially acceptable to use computers. J. Katz (2000)

24

demonstrated this in his case study of two young geeks from an isolated town in Idaho. Based upon their talent with computers, they moved to Chicago and ended up finding financial and social success that they would not have had otherwise. These socioaffective qualities combine with intellectual and creative abilities to form the natural foundation that an individual uses to develop his or her computer technology talent.

Besides natural abilities, there are also intrapersonal catalysts within these gifted individuals that affect their talent development. One of those catalysts is the quality of motivation, representing a person's needs, wants, and values. Some highly talented children find success at a young age and are very motivated to achieve their goals quickly (Machlowitz, 1985). It was internationally recognized by Van Lieshout and Heymans (2000) that, "Talented developmental pathways are assumed to be the result of goal-oriented individuals reciprocally and successfully interacting with changing contextual opportunities and constraints" (p. xv). Although there are many highly motivated gifted individuals, there are also those who choose not to achieve and struggle with underachievement (Winner, 1996; White, 2000). In the book *Smart Kids* by Durden and Tangherlini (1993), they examined the case study of one gifted boy who struggled academically and hated going to school. In sixth grade, his parents bought him a computer and he showed intense motivation while programming and working with his friends to plan their own computer company. Afterwards he turned those feelings of motivation towards his school work and easily became one of the top students in his class. According to Feldhusen (1995), "Highly talented youth must come to an awareness and understanding of their own talents and abilities. This they do chiefly through the challenging, accelerated, and enriched learning experiences" (p. 92). Having external

25

challenges can help gifted students enhance and develop their skills, and working with computers is one way for them to achieve both internal and external rewards.

Another quality related to intrapersonal catalysts is volition. According to Gagné (2003), the concept of volition contained the concepts of will-power, effort, and persistence. Students could have motivation for certain goals, but they also needed to put in continuous hard work in order to be successful. Feldhusen (2005) wrote that it was important to be realistic about the success rate for students with high potential: "The few who are willing to work very hard, who strive for expertise or for creative achievement, may go on to regional, national, or international recognition and make major contributions in the arts, sciences, and other fields" (p. 74). Near the beginning of the computer industry, Kidder (1981) conducted a year-long ethnographic study on a group of programmers constructing a new computer system. What fascinated Kidder was the intensity and commitment that these people showed by working day and night toward their goal, often to the point of physical exhaustion. It was hard for him to distinguish if these were ingrained personal characteristics or imposed external conditions demanded by the computer company, but he was impressed by their level of dedication. Resiliency was another characteristic related to this concept, and the ability to bounce back after suffering a setback could positively affect an individual's talent development (Horowitz, 1987; Jenkins-Friedman & Tollefson, 1991). It seemed that the field of computer science was particularly demanding of an individual's time and energy, but for certain individuals like Bill Gates and Steve Jobs, all of that effort eventually paid off with the success of their companies (Landrum, 1993).

26

The aspect of self-management contains the qualities of concentration, initiative, scheduling, and the work habits of talented individuals. Some of the sixteen habits of mind developed by Costa and Kallick (2000) that fall under this heading are: Managing impulsivity, striving for accuracy, taking responsible risks, and remaining open to continuous learning (pp. 2-13). Gifted students with high self-management and self-regulation were able to plan the steps necessary to achieve their goals and have confidence that they would successfully complete the task (Zimmerman & Martinez-Pons, 1990). These qualities of personal responsibility are helpful in the fast-paced world of technology. In Gagné's (1999) book on peer nominations for gifted programs, he presented a brief description of students with computer technology talent: "A programmer can learn new programs alone and does not need to ask for help when a program is not working" (p. 79). Many computer enthusiasts describe themselves as self-taught and take the initiative to acquire as much knowledge as possible while exploring on their own (Papert, 1993). Selwyn (2005) tried to determine the most effective way people learn computers, and he found that:

> Learning to use a computer is a complex, gradual process that takes place over the life course... Although sometimes punctuated and stimulated by formal learning, most computer learning is informal, fragmented, and specific to the individual's context of computer use at the time. (p. 130)

If most learning takes place informally, then it makes the developmental stages of computer science harder to predict. It is important for educators and parents to provide environments conducive to these self-directed learners, and give them the freedom to set their own pace and play with different approaches to learning (Rathunde, 1992).

The category of personality includes an individual's temperament, traits, well-being, self-awareness and esteem, and adaptability. This area has a large range of variability within the gifted child population, but theorists have tended to focus on how these individuals use these aspects to promote or hinder their own development (Friedman-Nimz & Skyba, in press). Davis (2003) described the positive personality characteristics related to creativity: Original, aware of creativeness, independent, risk-taking, motivated, curious, sense of humor, attracted to complexity, artistic, open-minded, needs time alone, and intuitive (p. 313). He also provided details for some negative personality traits that could also appear in creative individuals: Temperamental, indifferent to convention, questions authority, egocentric, stubborn, demanding, sarcastic, and absentminded (p. 313). A person could have a lot of these traits or just a few, and generalization should be avoided when trying to describe the gifted population. In early personality research, Dabrowski and Piechowski (1977) emphasized that gifted children could be overexcitable in five possibly overlapping areas: (a) psychomotor, (b) sensual, (c) intellectual, (d) imaginational, and (e) emotional. These overexcitabilities could have a positive or negative effect, depending on if the gifted children's needs get met in their environment. Another personality aspect frequently researched in the gifted field is the concept of perfectionism. Schultz and Delisle (2003) concluded that fears of failure arise because, "The emotional stakes are raised with each judgment by self, peers, teachers, or parents as perfectionism blossoms into a full garden of thorny perceptions, expectations, and anticipated successes" (p. 488). Piirto (1999) included personality attributes as the base of her pyramid model of talent development, many of which have already been mentioned above. She also included the concept of *androgyny*, a concept where a person demonstrates a mixture of personality traits considered *masculine* or *feminine*. How all of these personality traits specifically affect computer technology talent has yet to be shown

28

through systematic research. The intrapersonal qualities of motivation, volition, self-management, and personality combine into a unique pattern for every individual that affects the path of their talent development.

Environmental catalysts such as culture, people, and resources play a large part in an individual's talent development (Gagné, 2003). To lay the groundwork for future accomplishments, Gardner (1999) suggested that an environment should be created, "That constantly stretches the young person, so that triumph remains within grasp without being too easily achieved" (p. 121). Many theorists agree that the family environment is one of the strongest determiners of success in a talent area (Gowan, Khatena, & Torrance, 1979; Bloom, 1985; Feldhusen, 1996; Winner, 1996). Even though parental influence was important, the *Media in the Home 2000* report stated that, "Parents reported regularly supervising their children's use of television (88%) but only half reported regularly supervising the internet or video games (50% and 48%, respectively)" (Meszaros, 2004, p. 387). In the early 1990s, Giacquinta, Bauer, and Levin (1993) found that the majority of parents offered no computer assistance to their children because of unwillingness or lack of skill. Parents who did demonstrate an active interest in their child's computing exhibited, "Six important kinds of involvement: provisioning, goal setting, modeling, coaching, praising, and scaffolding" (p. 63). This meant they would provide the computer and software, establish a clear purpose for the use of the technology, model exemplary behavior, offer suggestions, feedback, and praise, and support their children until they were proficient and then allow them to continue on their own. However these positive parental behavior were not frequently observed, and usually the children were left completely alone with the computer. This conclusion was

29

supported by Attewell, Suazo-Garcia, and Battle (2003) with their own observation that,

"Children who do have a computer spend almost no time on this computer accompanied

by an adult" (p. 290). Jordan (2002) emphasized that the family culture greatly influenced

how an individual's technology perspective developed:

> Families make available certain kinds of media within the home but also provide
> notions about how and when to use the media... Much of this socialization does not
> occur through the explicit directives of parents. Instead, the interactions of family
> members subtly create patterned ways of thinking about and using the media. (p. 231)

According to this, even if there are no specific rules and goals set by the parents, their

attitudes and beliefs about computers and media play a large part in their children's

interest in technology.

Besides the home environment, the school setting was recognized as another

important external factor in talent development (Freeman, 2000; Gallagher, 2003). Gifted

students could benefit from formal education, if teachers were willing to educate

themselves about the fun and responsible ways to use computers in the classroom

(Heaney, 2003; West, 1997; Walker de Felix & Johnson, 1993). Another group of

researchers explored an interesting question: Do intellectually gifted children use

computers differently than other children? Using a software package called *The Factory*,

investigators observed children ages 10 and 11, and compared the behaviors of students

from an enriched gifted class to students from a general education class (Bowen, Shore,

& Cartwright, 1992). The intellectually gifted students using this computer program

planned their steps better beforehand, and they also had fewer trial and errors before

creating their final product. All of the students accomplished the same activity goal, but

the study documented that intellectually gifted students used the software quicker and

more efficiently. In relation to the learning styles of gifted students, Bulls and Riley

(1997) suggested that computers can accommodate both students who prefer self-directed individual work as well as students who prefer collaborative project-based learning. In a more recent study on new literacies Leu, et al. (2007) described how the same qualities manifested in intellectually gifted students while they searched the internet to answer specific questions. The high ability students demonstrated fewer false steps and were able to scan web pages quicker to decide if they were helpful. Many of these articles discussed the classroom application of technology, but did not directly talk about how to specifically teach students with CTT. These results hinted that the processing behaviors of these students could be overlooked if schools only use standard identification procedures for gifted services.

Both in and out of school, the influence of peers has a large effect on the talent development of young people (Schneider, 1987). In Holloway and Valentine's (2003) recent book, *Cyberkids*, a study was conducted in the UK on how peer groups interacted in the classroom with computers. Of the boys observed, there was a subset that the authors dubbed the "techno boys" whose identity was tied to their ability and interest in computers (p. 50). Kerr and Cohn (2001) described how sometimes male students who were overly interested in computers could be socially isolated and considered outcasts: "Smart boys who are seen by their age peers as nerds typically exhibit idiosyncratic talent in one area, such as chess, fencing, poetry, or computer programming" (p. 242). Although boys and girls with CTT could be shunned by other classmates, they were often able to form friendships with other students who shared their same level of interest in computers. Subrahmanyam, Greenfield, Kraut, and Gross (2002) came to this conclusion about teenagers and computer use: "Adolescence in the United States is typically characterized

31

by experimentation with social relationships and an expansion of peer groups.... Teens' use of the internet for this social experimentation appears consistent with their developmental needs" (p. 19). At home, Orleans and Laney (2000) observed that the children who had a computer that belonged solely to them were more likely to explore the system with friends, surfing the internet, sharing ideas, and collaborating on projects. The effect of these peer relationships specifically on CTT development need to be explored further.

The individual abilities and intrapersonal characteristics within a person combine with environmental factors to produce systematically developed skills and talents. This was not a foolproof equation with guaranteed results, and Gagné (2003) hypothesized that only the top 10% of gifted individuals were able to successfully realize their talents in a specific area. Even with the right combination of qualities and experiences, there is still the factor of chance, or *co-incidence*, which ultimately affects if a person is in the right place at the right time to successfully develop their talents. By examining the developmental path of individuals in the area of computer technology, a better understanding of this process can be achieved.

Developmental History of the Field.

The development of the computer technology field is reflected in the shift from mainframes to personal machines and the increased accessibility to computers and the internet for the novice user (Rainie & Packel, 2001). Morelock and Feldman (2003) described how important these advancements were to talent development: "Prodigious achievement only occurs within domains accessible to children, meaning they require little prerequisite knowledge and are both meaningful and attractive to children" (p. 462).

Because more children across the country had access to technology, they were able to

expand their knowledge and skills at an earlier age. Also as the knowledge about

computers dispersed through society, the cultural expectations changed as well.

The historical timeline of computer science has also gone from isolated

laboratories to widespread use in the United States (Abbate, 1999; Horrigan, 2000).

Csikszentmihalyi and Robinson (1986) described how this situation was essential because

talent could only be recognized by experts familiar with the established benchmarks and

guidelines of a domain:

> Talent cannot be observed except against the background of well-specified cultural
> expectations. Hence, it cannot be a personal trait or attribute but rather is a
> relationship between culturally defined opportunities for action and personal skill or
> capacities to act. (p. 264)

It is only within the last few decades that the field of computer technology has reached

the point where distinct levels of knowledge can be successfully evaluated by a team of

experts. To map the specific stages in the development of this field, Feldman's (1994)

universal-to-unique continuum is outlined below.

Universal-to-Unique Continuum.

Feldman (1994) hypothesized that each field of knowledge has different levels of

intellectual achievement. There are some knowledge domains that are specialized where

only individuals who choose to pursue that area of expertise will acquire the skills and

information specific to that area. In *Beyond Universals*, Feldman (1994) proposed a

theory to explain how an individual progresses in an area of expertise above and beyond

the average person: "As we move from universal achievements to types of knowledge

which fewer individuals will acquire, we begin to move along a continuum from

universal to uniquely organized domains of knowledge" (p. 23). The five levels of this universal-to-unique continuum are (1) Universal, (2) Cultural, (3) Discipline-based, (4) Idiosyncratic, and (5) Unique. These were described as stages because they were sequential and a person could not skip a stage of learning without having acquired the knowledge and skills from the previous one. However, Feldman recognized that for some individuals with exceptional talent, the process could be compacted and accelerated. Each level is described in detail below.

Universal knowledge is considered attainable by all developmentally normal children. Piaget's largest body of work (1952, 1972) focused on the developmental stages that all children experience from birth to adolescence. The universal stages of development acknowledged by Piaget (1972) and other subsequent stage theorists after him includes the sensorimotor, preoperational, concrete operational and formal operational stages. Human beings are expected to go through these steps of intellectual reasoning, regardless of context. In the cultural category, different domains of knowledge are more important to certain societies than others. Some skills like reading and writing are emphasized in American culture and people are expected to have a minimum level of competence in this area. But other skills like paddling canoes, making weapons, and identifying different plants and animals are skills central to other cultures that are not currently valued by Western culture. If a person is a member of a certain culture, they are expected to attain the knowledge and skills important to their society.

The next level of knowledge is labeled as discipline-based. This is where an individual follows a career path and works to attain mastery in an area that is at a rank not expected to be achieved by the general public. Certain levels of expertise can be evaluated at this point, such as novice, advanced beginner, competent, proficient, and expert (Dreyfus & Dreyfus, 1986). Theorists Ericsson and Smith (1991) and Glaser

(1996) have examined expertise in some specialized areas like chess, music and sports.

Feldman (1994) stated that a person who has reached the master level in a discipline may

end up filtering their perception of the world through this lens: "The meaning of the term

discipline implies that one's way of thinking, of organizing reality (or part of it) takes on

a distinctive and traditional form as a consequence of acquiring that discipline" (p. 32).

Specifically related to technology, Campbell, Brown, and DiBello (1992) conducted an

in-depth study of computer programmers as they learned a new language and documented

the stages they went through in developing their skills. They noted that the programmers

followed a normal pattern of thought processes used by anyone trying to master a subject,

but there were also some features unique to this domain. The meticulous level of detail

needed for programming was noted as a perceivable difference between a novice and

master, "We found that our professional programmers used learning strategies that

required them to take up the 'programmer's burden' of understanding what happens in

each line of Smalltalk code" (p. 269). The researchers concluded that there were seven

distinct stages of skill development in programming, but they admitted that their choice

of a particularly difficult programming language might have affected the results of the

study[1]. Once a talented individual has achieved the highest level possible in their

discipline, they still might hunger to explore even further.

At the idiosyncratic level, an expert in the field continues on to further develop

expertise in a specialized sub-area of that domain. Feldman (1994) was clear about the

---

[1] Developmental levels of Smalltalk programming: (1) Interacting with the visual interface, (2) Syntax rules and order of precedence, (3) Locating classes and methods in the hierarchy, (4) Class vs. instance distinction, (5) Model – Pane – Dispatcher, (6) Object-oriented design, and (7) Grandmaster level. From Campbell, Brown, & DiBello, (1992), (p. 286).

qualities he believed were necessary for individuals at this level: "The hallmark of these idiosyncratic achievements is the complementarity between a field of endeavor and a set of individual predispositions or talents" (p. 34). Thus a person needed to have an exceptional talent in a subject to excel in that specialty area. For most of the human population, the next level of unique thinking is a hypothetical concept that may only be seen once or twice a generation in any given domain. In this sense, it is similar to the pinnacle of Maslow's (1991) self-actualization theory. The difference is that theoretically people can strive towards self-actualization in a hope to one day achieve this level of growth, but with unique talent some people will never reach that point no matter how hard they work in a discipline area. Feldman (1994) acknowledged that few people will ever reach the unique stage where they have the capacity to completely change a domain of thinking: "Only a tiny number of novel thoughts are ever perceived as useful at all, even a tinier number of novel thoughts are useful for very long, and even a smaller number yet become continuing parts of an evolving field" (p. 47). Examples of this can be easily seen in the science field, such as Isaac Newton and Albert Einstein, but any domain has the potential to be radically changed by the unique contributions of certain individuals.

The domain of computer technology has progressed through this continuum, and it is the individuals who show the potential for achieving the idiosyncratic and unique stages that this project is most interested in studying. It was easiest to see unique contributions early on in the computer science field. The era of computer was started by self-proclaimed nerds, like Bill Gates at the Microsoft Corporation. According to Feldman, every field has a continuum of levels from universal to unique. People that think at a unique level have the potential to revolutionize their own field, and sometimes the entire culture. The changes that have come about due to the development of computer technology have had a large impact on American society as well as the global culture.

Historical and Cultural Trends.

When tracking the historical and cultural contributions related to computer technology, it is clear that the crucible for creation was Silicon Valley in California (Lee, Miller, Hancock, & Rowen, 2000). With the convergence of so many innovative entrepreneurs in one location, the advancements made in the field were extraordinary. Graham (2004) stated that since the 1970s the culture has entered a new era of the computer renaissance: "Over and over we see the same pattern. A new medium appears, and people are so excited about it that they explore most of its possibilities in the first couple generations. Hacking seems to be in this phase now" (p. 33). Without these historical and cultural developments it would be more difficult to discuss the idea of someone having distinguishing qualities of CTT.

Although human inventions have always spurred changes in society, it seems that the impact of the computer into everyday lives has been uniquely profound. Greenfield (1984) compared it to another invention that redefined American culture: "Just as the development of gasoline engines provided a tool for human physical activity, so the development of the computer provided a tool for human mental activity... Tools are not just added to human activity; they transform it" (p.153). Johnson (1997) coined the phrase *interface culture* to describe the changes that have taken place because of computers and technology. With the advent of the graphical web browser *Mosaic* in 1993, a permanent impact was made on the culture of the United States and the world

37

(Abbate, 1999). This change has occurred more rapidly than other shifts, going from only about 35% of adults being online in the spring of 1998, to close to 70% at the end of 2006 (*Pew Internet & American Life Project*, 2007). It is hard at the turn of this century to imagine what the next 100 years will entail, especially as technology improvements increase exponentially: "The paradigm shift is currently doubling every decade; technological progress in the 21st century will be equivalent to what would require on the order of 200 centuries" (Kurzweil, 2001, p. 153). A certain level of knowledge of computers is expected in the U.S. culture, and those who have not adapted to this technology are at a disadvantage when it comes to higher education and the work force.

To help prepare our future technological leaders, a major priority in this nation has been focused on the concept of computer literacy (Moursund, 1977). A problem develops when trying to define exactly what that means, at the national, state, and district levels. The International Society for Technology in Education (ISTE) (2005) and other organizations have outlined what it is that students should master at different levels, however these standards do not include differentiation for gifted and talented students. As Collis and Anderson (1994) explained: "Functioning with information technology requires certain knowledge, skills, and attitudes, the specifics of which depend upon one's environment and goals" (p.59). There has been a shift since the late 1990s to understand the development of an individual's concept of technology as a whole, instead of only being trained to use one program at a time. Tyler (1998) recognized this shift in thinking and described a proposed new model for thinking about computer literacy:

> This theory requires the removal of "application theory" from the focus of computer literacy training. In other words, a distinction is made between teaching someone to use a computer in general, and just teaching someone to use a bunch of computer programs. (p. 6)

38

A more integrated perspective in relation to technology literacy is permeating American culture as the next generations move into the workforce.

Howe and Strauss (2000) wrote the book *Millennials Rising* with information entirely about the generation born after 1982. They described this group of young people as optimistic, tech savvy, and focused on making the world a better place. Whereas older generations used computers and the internet for finding information, conducting business, and writing e-mails, this generation enjoyed using it for playing games, creating personalized spaces on social networks, and instant messaging (Fox & Madden, 2006). They have been labeled "Generation M" by Rideout, Roberts, and Foehr (2005) for their intense connection to media and technology. Sometimes it became difficult for these two groups to communicate effectively with each other. Chandler (2005) described how it was almost like they were speaking two different languages: "Both newcomers and insiders feel that members of the other group are illiterate in practices which they consider fundamental for effective communication" (p. 5). This conflict was not just between age groups, and there was a digital divide recognized among families with different levels of SES as well as among rural, suburban and inner-city students (Tyler-Wood, Cereijo, & Holcomb, 2001). Not everyone born in the 1990s spoke the language of technology, but its impact has been felt by the majority of children that age (Lenhart, Lewis, & Rainie, 2001).

Much has been researched about the general impact of technology on children and how their lives have been affected by growing up in the digital age (Lawler, 1985; Stonier & Conlin, 1985; Gill, 1996; Hellenga, 2002; Subrahmanyam, Greenfield, Kraut, & Gross, 2002). As with most new technologies, people took extreme positions in predicting both positive and negative outcomes for children in this society. In 1980 Papert touted his programming language for kids, LOGO, as a learning tool that would revolutionize the way math was taught in all elementary schools. He showed success with a small group of intellectually gifted students, but was unable to reproduce his positive results on a larger scale. Stoll (2000) was on the opposite side, vehemently arguing that computers should be kept away from young children. He stated that sitting and staring at a tiny screen would stunt their social, intellectual, and physical growth. Finding a realistic balance between these two perspectives has been a challenge for parents and teachers alike when dealing with children and computers. From a more optimistic perspective, Campbell (1990) stated that it was an exciting time for educational technology research because results could quickly find success in modern classrooms:

> When we study children learning basic mathematics, we are dealing with a field that changes slowly, and our findings are unlikely to change it. When we study children using computer technology, we are dealing with a field that changes rapidly, and our findings can help to guide that change. (p. 382)

A multitude of studies have been done on the impact of technology in the classroom and how computers affect learning at all ages, both positively and negatively (Armstrong & Casement, 2000; Cuban, 2001; Gow, 2004; Oppenheimer, 2003; Plowman & Stephen, 2003). The policies of the United States government have changed to reflect the growing need for computer technology resources in the public education system. A National Technology Plan was recently developed to examine, "The success of innovative new approaches to learning through advances in educational technology" (U.S. Department of Education, 2005). Researchers continue to investigate the impact of technology on children's lives in a variety of settings.

Many children have become authorities in the area of technology before their parents, and, "The digital revolution, unlike previous ones, is not controlled by only adults" (Tapscott, 1999). As society continues to evolve, the duty of parents to ensure their child has "technology literacy" becomes more critical, especially if the child has demonstrated talent in that area (Siegle, 2004). This involves not only knowledge of computers but also the ethical issues that come up when interacting with others in an on-line environment, such as flaming, stealing information, and cyberstalking (Schwartau, 2001). An extensive study by Holloway and Valentine (2003) concluded that:

> Overall, most children use ICT in a balanced way. The technology is fitted into their lives rather than displacing other activities; as such it does not make them narrowly home centered, nor erode their use of outdoor space. Rather, its use is taken for granted as an unremarkable part of everyday life - a finding replicated by other research. (p. 123)

For most children and adolescences, computers are mainly used for fun (Rushkoff, 1999; Strot, 1999; Smith & Grant, 2000; Gee, 2003). The entertainment value of technology in our culture was described by Bryce (2001): "Technological leisure activities fulfill the same functions as those considered traditional; they provide relaxation, stimulation, escape, social interaction, and the development of self-identity and lifestyle" (p. 9). Johnson (2005) stated the opinion that computers and video games have actually raised the general intelligence level of society because complex thought and interactions are required while playing with these devices. Besides the entertainment value of technology, it has also made a large impact on the field of education.

The historical and cultural impact of computer technology has been profound. From the types of jobs people do to the forms of leisure they enjoy, the effects of the digital revolution can be seen everywhere. The impact on the learning of the younger generations is still being assessed, and whether these changes will ultimately have a positive or negative effect is unknown. However, schools should accurately reflect the needs of the culture, and currently those needs include computer technology.

Evolutionary Time.

Feldman and Benjamin (1986) described his last factor as the overarching umbrella under which everything else falls: "Evolutionary forces operate on individuals through the biological processes of variation and natural selection, and on the accumulated products of collective efforts – that is, cultures and their artifacts – through more macroscopic processes" (p. 14). This aspect is important to acknowledge, although they admit that the evolution of the brain structure takes hundreds of years and cannot be measured in short-term research. Csikszentmihalyi, Rathunde, and Whalen (1993) put forth the idea that the purpose of evolution is to further those individuals in a society that have the most to offer:

> After a society takes care of what is needed to continue physical existence, the next most useful goal in which to invest resources is the development of individual gifts. A group that followed these priorities would be in the best position to evolve further, and it would face the future with the best chance for success. (p. 28)

According to this opinion, the development of talents such as CTT should be encouraged for the betterment of the entire human race.

There has been an interest in the field of psychology to discern the effect that computers have had on human cognition. For example, one study hypothesized that there were two key cognitive effects of technology on an individual: increased performance while using technology, and positive residual effects after using technology (Salomon, Perkins, & Globerson, 1991). Salomon et al. concluded that computers did make a noticeable difference in human thinking in their sample, but this was determined on an

42

individual basis. Clark (2004), author of *Natural Born Cyborgs*, wrote that the brain is more than capable to expand as technology continues to progress: "Our self-image as a species should not be that of ancient biological minds in colorful young technological clothes. Instead, ours are chameleon minds, factory-primed to merge with what they find and with what they themselves create" (p. 141). Unfortunately, Clark's book was based mostly on theory and abstract predictions, and discernable changes in the brain have yet to be demonstrated.

Some attention has been paid to the brain structure and thought processes of the "Net generation" who appear to be qualitatively different from other generations. Prensky (2001a) coined the term *digital native* to describe the children who grew up with computers and video games. In a follow up article, Prensky (2001b) quoted information from the 1997 *Inferential Focus Briefing* that stated, "Children raised with the computer think differently from the rest of us. They develop hypertext minds. They leap around. It's as though their cognitive structures were parallel, not sequential" (p. 4). This concept of a hypertext mind with a web-like structure like links on the internet intrigues scholars. However, tests to distinguish this type of thinking have yet to be successfully developed. In a *Time* magazine article by Wallis (2006), the psychologist David Meyer studied the brains of children as they multi-tasked using MRI scans. He concluded that, "Habitual multitasking may condition their brain to an overexcited state, making it difficult to focus even when they want to" (p. 53). Although it appears that young people are more adept at branching their cognitive processes, other aspects of attention and reflection can suffer. The long-term effects on brain functioning are being studied, but it may take some time to see the final results of too much interaction with technology.

Promoting human talent is a necessity for any society to grow and evolve. In American society, technology affects how people think and behave, and the long-term impact will not be known for many generations. Any deeper analysis into the evolution of the human brain is beyond the scope of this study.

Summary

Feldman (1994) stated in his co-incidence theory that four conditions have to align to create an environment where prodigious youth can develop extraordinary talents. In the domain of computer technology, historical and cultural forces have produced a window of opportunity for the next generation of geeks to demonstrate their skills. This study focused primarily on (1) the individual's life span, and the personal qualities and external influences of the participants. The potential impact on the (2) development of the computer technology field, (3) historical and cultural trends, and (4) the evolution of the human race are discussed briefly in chapter 5.

CHAPTER 3

Methodology

To get a better idea of the development of computer technology talent (CTT), the personal backgrounds and experiences of three samples and a known group were explored. Interviews with notable adult programmers conducted in the 1980s were analyzed to discover patterns of personal, cognitive and affective qualities related to their work. They were also asked about the internal and external forces that affected the development of their area of expertise. These were compared to interviews with three real time samples, representing modern technology workers, college students, and high school students, to see if there were similarities in the CTT development among these four groups. An inductive approach was used to gather pertinent information on the life experiences and personal characteristics of all of the participants in this study. The developmental process was examined in order to gain a more complete understanding of the attributes that define this area of expertise.

Research Objectives

The research objectives of the proposed study were: (1) to describe in rich detail the distinguishing intellectual and personal qualities of adults and adolescents demonstrating computer technology talent; and (2) to trace the developmental path of computer technology talent in the lives of adults and adolescents.

Research Question.

1. What cognitive and affective qualities and life events mark the development of computer technology talent (CTT)?

Qualitative Design

Due to the emergent nature of the research question, a qualitative design was specifically chosen for this study. I wanted to collect personal narratives in which the participants could describe their own memories and express their cognitive and affective qualities (abilities, beliefs, and behaviors) that influenced their computer interests. Qualitative inquiry has proved valuable in the field of gifted education, as discussed by Cross (1994): "Approaches and techniques underpinned by research assumptions that are not positivistic in nature are showing great promise for the future for the identification of children [in gifted education]" (p. 285). In a key article, Coleman (2001) successfully used ethnographic methods to detail the lives of a group of gifted high school students, weaving information together artistically like a 'rag quilt.' It is no longer necessary to justify using non-empirical methods for the basis of educational research.

The theorists Lindlof and Taylor (2002) promoted research that builds from the bottom-up: "Most often in qualitative research, a strong current of inductive thinking stimulates the development of categories; that is, a category begins to form only after the analyst has figured out a meaningful way to configure the data" (p. 215). Instead of starting with a preconceived structure, the theory of giftedness most suited to this study only became clear after the data analysis. It was important to me that the categories needed to come from the results of interactions and interviews with the people themselves, without predetermined assumptions. A cursory exploration of research on qualities of gifted individuals showed samples almost exclusively selected based on their intellectual abilities, and second on their personal qualities. Renzulli and Purcell (1996) saw this as a flaw in gifted research and encouraged a different approach: "Giftedness is

46

not necessarily an inborn and enduring trait as previously thought, but one that emerges in some people, in some areas, and under certain circumstances" (p. 175). It is under this new perspective that this current study was conducted.

Within the field of gifted education, there are three main modes of inquiry that have been used by researchers in the past. In an article by Coleman, Sanders and Cross (1997), they referred to these research types as empirical-analytic, transformative, and interpretivist. The approach used in this study falls under the heading of interpretivist, which is where, "Knowledge is viewed as subjective and what can be learned is how others understand the world. Interpretivist researchers work to uncover the patterns or rules in social relationships as seen by groups and individuals" (p. 107). This research perspective is relevant because it emphasized that giftedness can mean different things to different people depending on the context.

Rationale for Samples

When this study was originally conceived, only one contemporary sample was going to be collected from the population of high school students in the area. But due to the evolving nature of a qualitative design, it soon became apparent to me that this phenomenon would be better analyzed as a comparison study. The experiences of the individuals in the historical group (Sample 1) were different from the contemporary group (Sample 4), but having only those two data sets did not adequately capture the developing nature of computer technology talent. Initially I contacted the individuals in the snapshot group (Sample 2) on a purely consultative role to get their opinion on my research into CTT, but I found that they had much more to offer on the subject. They became so involved in the study that I decided to recruit them as participants and use

47

snowball sampling to interview other people that they recommended in the same peer group. Finally, I was fortunate to have access to the longitudinal group (Sample 3) based on previous research conducted by my university advisor. I was able to contact them to do follow-up interviews about their experiences with computers and see if their involvement had changed from high school to college. With these four samples the development of CTT could be traced from the 1980s to 2007, allowing similarities and differences to be examined along this timeline.

Sample 1: Historical

   Setting.

The interviews for this sample were collected from an original primary document that contained unannotated data. Over 18 years ago, a book was published by Microsoft Press titled *Programmers at Work*, compiled by Lammers (1989) who was the Associate Publisher and Director of Multimedia Publishing at the Microsoft Corporation at that time. She interviewed some of the founders of the modern field of computer science who were located in different parts of the country, and there was even one programmer from Japan. At the time, Berkeley and Silicon Valley in California were the epicenter for computer programming, but there was also a strong presence on the East Coast with Massachusetts Institute of Technology (MIT) and businesses like International Business Machines Corporation (IBM) and Lotus Software. In her interviews, questions tended to focus on the programmers' current projects, and their perspectives on the positives and negatives of the computer business. Lammers (1989) wanted this book to present "The experiences, approaches, and philosophies of software designers in a personal, in-depth manner" (p. 1). These interviews were compiled and published in a completed transcript

48

format, with very little analysis except for an introduction section and a brief biography of each programmer. In order to benefit this particular study, I focused on the intellectual, social, and intrapersonal patterns of these programmers, as well as the childhood experiences that had influenced their love for computers.

Included in this sample was one other interview conducted separately by Subotnik (1993) for the *Journal for the Education of the Gifted*. In its series of interviews called "Conversations with Masters of the Arts and Sciences," Subotnik interviewed Joseph Bates, a math prodigy, who received both his bachelor's and master's degrees in computer science by the time he was 17-years-old. He was the first student to be identified and nurtured by famed gifted researcher and educator Julian Stanley at Johns Hopkins University. This interview was included because questions were asked that specifically addressed the area of gifted education and computer science, and this interview had also been published in a transcript format with no additional analysis by the author. By starting with these adult programmers, insight into the original characteristics of this group helped lay the groundwork for the interview structure of the other samples.

Participants.

The names of the adult participants (N=20, Male=20) are recognized in the computing field and are used in this study because the interview data has already been published in a publicly available format. This known group served as a baseline for the comparison with the other three samples. In choosing the people she wanted to interview, Lammers (1989) desired to get a cross-section of ages, experiences and specialties from different areas of the software industry. The individuals she chose were emiment and had all made a significant contribution to the field in some way. The first eight programmers had a West Coast connection and included Charles Simonyi, Butler Lampson, John Warnock, Gary Kildall, Bill Gates, John Page, C. Wayne Ratliff, and Peter Roizen. These participants had different levels of experience at Xerox PARC, Berkeley and Microsoft, as well as starting their own independent companies.

The next five people, Dan Bricklin, Bob Frankston, Jonathan Sachs, Roy Ozzie, and Bob Carr, had East Coast connections to MIT, IBM and Lotus Software. Two of the principal creators of the Macintosh were also interviewed, Jef Raskin and Andy Hertzfeld. Lammers (1989) pointed out that the last four she spoke to were non-traditional programmers who chose to focus more on music and graphics. This included Scott Kim, Jaron Lanier, Michael Hawley, and Toru Iwatani from Japan. The interview conducted by Subotnik (1993) with Joseph Bates provided details of his current position as a professor of Computer Science at Carnegie Mellon University. It is assumed that all ethical procedures were followed by Lammers and by Subotnik when they conducted their interviews.

Sample 2: Snapshot

Setting.

For this sample, I wanted to find participants who were close to my own age and could be considered my contemporaries. I was already very familiar with my own experiences growing up with computers, and I knew that my personal development was going to affect how I analyzed the results of this study. A snapshot sample was taken to examine one point in time in the adult lives of these technology workers. The participants were from different locations all over the country, from Seattle, WA, to Washington, DC. There was even one participant who grew up in the Midwest who moved to Japan to continue his studies. Even thought they were coming from very different geographical backgrounds, I wanted to investigate if they all had a similar experience growing up with computer technology.

Personal computers were still fairly new to households in the 1980s, reflected in individuals' reports of variability in exposure to technology while growing up in this sample. Almost all of the participants were still adolescents when the World Wide Web went public in 1993. They were part of the generation that came of age during the digital revolution of the 1990s (Fox & Madden, 2006). In particular, I was interested to determine how this generation was affected by the historical and cultural events at this time, and if this affected their development of computer technology talent.

Participants.

All the participants (N=9, Male=9) in this sample were currently or recently employed in jobs related to computer technology. The majority of them received gifted services at some point in their educational careers, which supported their inclusion in this study of talented individuals. They were all in their late twenties or early thirties at the time of interview, and had a variety of experiences with technology, both positive and negative.

I used snowball sampling to recruit many of these subjects. Participants were asked if they knew other people in their industry who shared their same interest and level of proficiency with computers. Through different online networks, contact was made with these individuals and permission obtained for inclusion in this study. They agreed to answer interview questions through e-mail, live chat or face-to-face meetings, depending on their location (Appendix A). Many were interested in the objectives of this study, and offered their own opinions about the development of computer technology talent.

Sample 3: Longitudinal

    Setting.

        The original conception for this research study started with a pilot study of gifted students, detailed in the published article by O'Brien, Friedman-Nimz, Lacey, and Denson (2005). The focus of the pilot study was to explore the formative experiences, cognitive abilities, and personality characteristics that could be labeled as part of computer technology talent (CTT). Findings included discernable patterns and recurring themes in these students' lives, reflected in their histories with computing, family support, and key educational experiences. Results of this pilot project suggest that there were two sub-types of computer technology talent among these adolescents: programmers and interfacers (Appendix B). After the results of the pilot study, I wanted to continue this research and expand the sample to include more people from a variety of settings. I also wanted to follow-up on this particular sample to examine their continued development in the area of computer technology. O'Tuel (1991) described how longitudinal studies have provided valuable data in the field of gifted education, and were important when examining the aspects of development. Therefore, a longitudinal approach was taken to compare the original data and follow-up interviews conducted four years later.

        The original high school student interviews were conducted in a medium-sized Midwestern college town (population approximately 80,000 full time residents). At that time, slightly more than 10,000 students were enrolled in the public school system

(Friedman-Nimz, Lacey, & Denson, 2002). The local school district featured two high schools, four junior highs and eighteen elementary schools, as well as one accredited virtual school that served K-12. The district had a clear process established for the participation of human subjects in research, including obtaining permission from each participating school's principal. For the follow-up interviews, the original pilot study students were contacted electronically by the researcher. They were all attending different colleges and universities, in various locations from the east to the west coast.

Participants.

The pilot study participants (N=9, Male=8, Female=1) were juniors and seniors in high school who all demonstrated computer technology talent. The project was open to all students in the computer programming club, but only the older students self-selected on the basis of their interest in the project. After volunteering, these nine members participated in a Learning Generation technology-based project with the researchers[2]. As part of the project evaluation, each student took part in a structured interview with one of the researchers (Appendix C). These questions encompassed perceptions and evaluations of the project as well as explorations of dimensions of computer technology talent. They also completed a rating scale related to their thinking styles (Appendix D). This initial data was brought back for further analysis under the current research objectives and structure.

Four years after the initial study, an attempt was made to contact these students through the social networking website, Facebook.com. This site allows college students to connect with their classmates and post information about their current activities. Eight

---

[2] The development and evaluation of the Learning Generation (LearnGen) model was supported in part by an award from the United States Department of Education Preparing Tomorrow's Teachers to Use Technology initiative (P342A-990271-00).

53

students were successfully located this way, and seven agreed to participate in the current study (N=7, Male=6, Female=1). Follow-up interviews were conducted through e-mail with an abbreviated list of questions (Appendix E). The focus for this sample was to see if their interest in computer technology changed from the first data collection point, and to find out what sort of careers they hoped to pursue. They were also asked their opinions about the importance of computer technology talent to the field of gifted education, and what salient qualities this type of student might exhibit.

Sample 4: Contemporary

Setting.

For the final sample, the researchers returned to the same school district where the original pilot study was conducted. The school population had not changed, with roughly 10,000 students enrolled in the public school system (*Organization Statistics*, 2007). This school district encompasses the residents of a medium-sized Midwestern college town with a range of SES. In the ensuing time since Sample 3 attended public school, technology improvements had been made in the high schools. New computers and software were installed and made available to the students in this sample.

At the high school where all but one of the students attended, there was an independent study group (referred to as Phoenix Films) that created videos of different events for the school. This film media group would attend sports events, homecoming dances, and other school functions in order to document the highlights on video. There was a computer lab set aside especially for them where they used digital cameras and editing software to produce DVDs for the school. The members of this film media group were interviewed about their talent creating movies using computers, and asked how they developed their skills. Also, gifted students not in this media group were interviewed separately to determine their opinions about the computing resources at their school.

Participants.

Students in this sample were adolescents (N=8, Male=8) who demonstrated high interest in computer technology, and whose teachers indicated would be good subjects for this study. Most students were also identified as intellectually gifted according to Kansas state policies; however, this was not a label that determined the subject's inclusion or exclusion from the study. A variety of experiences were found, as there were participants from tenth to twelfth grade.

A triangulation of data was collected from the students, teachers, and parents for this sample. By having three separate sources of data, a rich description of this student group was developed. Seven parents were contacted, and three agreed to participate in this study. The parents (N=3, Male=2, Female=1) of the interviewed students were asked questions about technology in their home environment (Appendix F). They were contacted by phone or e-mail to establish contact and give their assent for participation. The pertinent teachers (N=5, Female=5) were also interviewed using questions geared towards them as educators (Appendix G). All of the teachers had direct contact with the students in this sample, either through gifted services or teaching in the media department.

For the contemporary sample, two focus group interviews were conducted with the film media group and with a group of gifted students. Focus groups refer to a "Nondirective technique that results in the controlled production of a discussion of a group of people" (Flores & Alonso, 1995, p. 84). The purpose of these group interviews was to observe the interaction between the students and their peers, and have a conversation about their technology interests. The semi-structured interview used for the focus groups centered on their school activities and collaborative projects they had worked on (Appendix H). These interviews were audio taped and transcribed by the lead researcher, and all identifying information was removed.

Data Collection

Interviews for Sample 1 were taken from the published book, *Programmers at Work* (1989) and the individuals were not contacted personally. For Sample 2, they were contacted electronically and responded to interview questions either through chat or e-mail. Information from Sample 3 had been previously collected during the pilot study using audio tapes and the researcher's notes. For the follow-up interviews, they were contacted electronically and responded through e-mail. The subjects for Sample 4 were recruited by contacting the pertinent gifted education teachers of each participating high school. The teachers were asked to recommend students who they thought demonstrated computer technology talent in their classrooms based on the thumbnail sketches developed from the pilot study (Appendix B). Once students had been identified by the teachers, an introduction sheet was given to them that briefly described the project so they could decide to participate or not (Appendix I).

Permission slips were distributed to the students in Sample 4 to get their signature of assent, and parents were asked to sign the informed consent as well (Appendix J). Because it was important to the researcher to not ask the participating teachers to do any extra work in their classrooms, the permission slips had self-addressed stamped envelopes attached to them so that they could be mailed back to the researcher before beginning the interview process. This was a risk taken during recruitment, because it relied upon the students to take the initiative in completing the permission form and putting it into the mailbox in order to participate. There were 70 permission slips distributed to students in an advanced science course and to students indentified as gifted by their school district. Three were returned through the mail and four were returned to the researcher in person. The percentage of returns was low (10%), however this number was representative of the percentage of gifted students thought to be in the normal student population (Gagné, 2003). Those who agreed to participate reflected the limited number of students in this area of gifted education.

Semi-structured interview questions were used to collect the data for this study. An outline was developed with specific starter questions, but flexibility was allowed for follow-up probes and prompts (Arskey & Knight, 1999). Not only was I interested in the respondents' opinions on their current use of computers, but I also wanted to know more about what circumstances and events in their lives led to their technology interest. Questions were asked about their predictions for the future to get their insight into the development of the computer technology field. Often the interviews turned into conversations and other related topics were brought up. Fontana and Frey (2003) emphasized that flexibility and spontaneity are positive aspects: "Interviewing and interviewers must necessarily be creative, forget how-to rules, and adapt themselves to the ever-changing situations the face" (p. 80). I preferred this method to a fully structured interview because it allowed for unexpected responses to emerge.

The majority of subjects who agreed to participate were interviewed individually and privately, using an on-line chat format or e-mail exchanges. The decision was made to employ digital communication for several reasons. First, an immediate record was made of the conversation, and the step of transcribing from audio tape was eliminated. Second, chatting on-line was the preferred way of communication for most of these individuals, and they were more comfortable sitting at home and responding through a keyboard than through a microphone. Third, the textual nature of chat and e-mail allowed for more reflection time, and the students could think about what they wanted to say before they typed it out. This method of data collection was successful in obtaining useful information from the respondents.

There were positive and negative aspects of using computer mediated communication (CMC), as outlined by Fontana and Frey (2003): "Face-to-face interaction is eliminated, as is the possibility, for both interviewer and respondent, of reading nonverbal behavior or of cuing from gender, race, age, class, and other personal characteristics" (p. 97). The potential for the respondents to have bias related to age, race etc. was lessened because they could not see the interviewer. However, all helpful nonverbal signals were removed, such as the subjects shifting in their seats to show uncomfortableness, or smiling and laughing to indicate happiness. In an on-line environment, these signals are reduced to symbols, called emoticons, to express emotions (e.g. :) for a smile). The researcher needed to be well versed in the nomenclature of online speak to successfully communicate with the subjects in this environment.

Validity and Reliability

The term validity is more closely associated with quantitative research, as the results of qualitative research are not expected to be completely objective. These interviews involved participants in real life situations, talking about their personal experiences growing up with technology. Smith and Deemer (2003) pointed out, "We cannot adopt a God's-eye point of view; all we can have are 'the various points of view of actual persons reflecting various interests and purposes that their descriptions and theories subserve'" (p. 431). The only thing that the researcher can hope for is that the facts will be presented in a clear fashion to the audience. Eisner and Peshkin (1990) stressed that, "Valid interpretations and conclusions function as surrogates through which readers of research reports can know a situation they have not experienced directly" (p. 97). If the development of computer technology talent is laid out truthfully, then the interpretations of the researcher and the reader should be similar.

Another concern for this study was ensuring that the interviews were reliable from sample to sample, and participant to participant. Silverman (2005) defined reliability as, "The degree of consistency with which instances are assigned to the same category by different observers or by the same observer on different occasions" (p. 210). There was only a single researcher involved in the analysis procedure, but the data were reviewed multiple times and revised each time new data were collected. Lindlof and Taylor (2002) offered further suggestions to strengthen the results of research studies: "In qualitative inquiry, validation can be achieved by evaluating multiple forms of evidence (triangulation and disjuncture) and by cycling some of the accounts back through the participants (member validation)" (p. 240). Both of these procedures were employed to increase the validity and reliability of this study. Further details are discussed in the conclusion.

Analysis Procedure

With the on-line interviews, the entirety of the data was contained in the chat history and e-mail messages. The focus group and teacher interviews were tape recorded and transcribed by the lead researcher. Having the exact words and comments that the participants made was essential, as well as the way the questions and probes were phrased by the interviewer. The resulting transcripts were open coded, a process that involves "the initial, unrestricted coding of data" (Lindlof & Taylor, 2002, p. 219). This step in the analysis focused on the patterns of life experiences and personal characterstics, both within and between subjects. To facilitate this coding, the qualitative data analysis software N6 (NUD*IST 6.0) (2002) was used to keep track of unit relationships and to produce graphical representations of all of the code types. When coding was complete, categories were defined that encompassed the shared qualities of different coded material. Salient quotes from each sample were compiled, and any dissenting opinions were examined as well.

Because of the high ability of gifted students to reflect on their experiences, a key step in the analysis process was to provide a summary of the results and allow the participants to comment on how closely it matched their experiences. Coleman, Guo, and Dabbs (2007) supported this idea that this group of people can provide a unique insight: "Gifted persons construct meaning and reinterpret meaning that others do not see. It is the content of the mind, the meaning itself, that is the origin of giftedness and creative productivity" (p. 52). It was a goal of mine that the students, parents and teachers would feel more like collaborators and would know that their input was valued and essential. The member check process, otherwise known as participant verification, was used to verify with the participants that the conclusions gathered from their data actually reflected their beliefs and opinions (Rossman & Rallis, 2003). The preferred method was

60

through e-mailed documents, and contact was maintained with the subjects during the course of the study for the purpose of debriefing.

Ethics

There was no predicted physical or emotional harm that came out of participation in this study. The participants were informed that there were no risks associated with participating in this project. To preserve the privacy of the subjects, pseudonyms were applied to the students who were individually interviewed. In relation to the computing field, the researcher chose to represent each person with a pseudonym taken from a different computer language to show the variety of programs created (*Scriptol.org*, 2007). When further details needed to be altered, (e.g. name of school, name of clubs) I did so in order to protect the identity of the students involved. Full disclosure was used in all aspects of the study, and participants were free to ask questions related to the purpose and publication of the final data. The participants could refuse to answer any interview question, and they had the option to withdraw from the study at any time.

Summary

The research presented in this study is based on qualitative interviews almost entirely collected using computer mediated communication (CMC). Interview questions were geared towards uncovering the intellectual, creative, and affective qualities that manifest themselves in relation to the development of computer technology talent (CTT). To provide both depth and breadth on the topic of CTT, four samples were taken from historical, longitudinal, and contemporary sources. In total, 61 participant interviews were obtained and over 10 hours of e-mails, chat transcripts, and tape recordings were analyzed. The resulting data for each sample are presented in Chapter 4.

CHAPTER 4

Interview Data

Because I was most interested in the phenomenological development of computer

technology talent (CTT), I began my investigation of the interview data free from the

constraints of preexisting categories. Each interview was examined individually and the

knowledge, skills, personal qualities, and experiences discussed by the participants were

compiled into a master list. After careful analysis of all of the interview data, the results

appeared to most closely align with Gagné's (2003) Differentiated Model of Giftedness

and Talent (DMGT). Although this structure was not initially used as a guiding theory for

this study, Gagné's model was the "best fit" for the emergent themes and covered all of

the qualities mentioned by the participants.

The first category in the DMGT model is natural abilities, which includes

intellectual, creative, and socioaffective abilities. The development of these natural

abilities is affected by two different types of catalysts. Gagné (2003) explained why he

chose this term: "A term from chemistry, *catalyst* designates chemical substances

introduced into a chemical reaction usually to accelerate it. At the end, these contributors

regain their initial state" (p. 64). The intrapersonal catalysts included motivation, volition,

self-management, and personality. The environmental catalysts explored in this study

centered on family, school, peers, and popular media. Finally, the different computer

technology activities where these groups demonstrated their talent were examined in a

variety of contexts.

The interview data for each of the four samples are presented below according to

Gagné's DMGT theory. The categories are: (1) natural abilities, (2) intrapersonal catalysts, (3) environmental catalysts, and (4) talent. A final section was included in the analysis of each sample related to (5) their predictions for the future of technology and society. This information provided insight into the development of the computer science field from the unique perspective of each group, relating to the second part of Feldman's co-incidence theory (Morelock & Feldman, 2003). Also a large emphasis was placed on future predictions in Lammers's (1989) original interviews, therefore I wanted to see what the other samples had to say on the subject. In this chapter each sample is analyzed individually and cross-sample trends are discussed in Chapter 5.

Sample 1: Historical

The participants in this sample (N=20, Male=20) were members of a known group of notable computer programmers. They offered examples from their lives and used analogies to illustrate their development in relation to computer technology. This early collection of interviews by Lammers (1989) focused on the distinctive characteristics of these noteworthy adults in the context of the computing industry. Examining the lives of these eminent programmers was helpful, because they were able to pursue their passion and achieved success in their chosen field. They also discussed external influences in their lives, from family to peers, and made predictions for the future of computing and the next generation of programmers. It was unknown if any participants were officially identified as gifted. Their age at time of interview, company location, and notable achievements are presented in Table 1.

Table 1.

Information Table for Sample 1

Name	Achievements	Age	Location
Charles Simonyi	Produced Microsoft Word and Excel	37	Microsoft, CA
Butler Lampson	Developed LISP and Mesa programs, computer professor at UC Berkeley	42	Xerox PARC, CA
John Warnock	Formed Adobe Systems for graphical design, developed PostScript software	45	Adobe Systems, CA
Gary Kildall	Developed first operating system, CP/M, published using CD ROMs	43	Digital Research, CA
Bill Gates	Founder of Microsoft, developed BASIC programming language	30	Microsoft, WA
John Page	Developed PFS:FILE software, and Image Database Management System	41	Hewlett-Packard, CA
C. Wayne Ratliff	Developed dBASE program, Worked for NASA Viking Flight Team	39	Ashton-Tate, CA
Dan Bricklin	Developed VisiCalc, first spreadsheet program, founder of Software Arts	34	Software Arts, MA
Bob Frankston	Developed VisiCalc, first spreadsheet program, founder of Software Arts	36	Software Arts, MA
Jonathan Sachs	Programmed Lotus 1-2-3 spreadsheet	38	Lotus Software, MA
Ray Ozzie	Developed Symphony program for Lotus	30	Lotus Software, MA
Peter Roizen	Developed T/Maker spreadsheet program	39	T/Maker, CA
Bob Carr	Chief scientist at Ashton-Tate, developed Framework program	29	Ashton-Tate, CA
Jef Raskin	Human-computer interface expert, started the Macintosh project for Apple	46	Apple Co., CA
Andy Hertzfeld	Lead developer and creator of original Macintosh Operating System	36	Apple Co., CA
Toru Iwatani	Developed first Pac-Man video game	34	NAMCO, Japan

Table 1 – *Continued.*

Scott Kim	Graphic designer, developed Inversions	34	Xerox PARC, CA
Jaron Lanier	Pioneer of Virtual Reality (VR) technology	25	Visual Programming Languages, CA
Michael Hawley	Developed software for the Sound Droid, used to edit music & audio files for films	24	Lucasfilm, CA
Joseph Bates	Completed computer science master's degree at Johns Hopkins by age 17	36	Carnegie Mellon, PA

Natural Abilities

Intellectual.

All of the programmers interviewed in this sample were considered highly intelligent, but they found different ways to describe their intellectual abilities. The interviews were casual and therefore Lammers did not collect any data pertaining to their academic achievements or scores on standardized tests to support this assumption of intelligence. When asked about computer technology talent, Bill Gates stated simply, "It's a talent, you bet. It's kind of like pure IQ" (p. 83). He went on to explain that some people just had the type of brain for programming, and that he would only hire talented people who could keep up with his own superior intellect.

Another aspect mentioned by this sample was the importance of having an excellent memory. A story related by Charles Simonyi showed his incredible memory capacity: "I had some aptitude when I was young. Even when I didn't know programming, I knew things that related a lot to programming. It was easy for me to remember complex things… When I was young, I could imagine a castle with twenty

65

rooms with each room having ten different objects in it" (p. 18). He then described how he used that extraordinary memory in his work to imagine a computer system with different software components in it. However, having such an extraordinary memory is very rare, so companies could not base their entire operations on such unique individuals. Peter Roizen, "A good memory for detail is important when I read my code, although I don't think I have a particularly good memory overall. I can't remember people's birthdays" (p. 196). Butler Lampson related from his own experience that, "Some people are good programmers because they can handle many more details than most people. But there are a lot of disadvantages in selecting programmers for that reason – it can result in programs that no one else can maintain" (p. 34). This statement shows how memory is admired, but other abilities need to work in concert to have a well-rounded computer programmer.

The rate at which a person could read code also seemed very important to most people interviewed in this sample. They were always trying to find faster and more efficient ways to write computer programs within their businesses. Speed was of utmost importance as a sign of talent, as Peter Roizen related, "I can pick up a piece of code I wrote a couple of years ago and generally quickly grasp what I was doing" (p. 203). Some of them talked about sometimes using code reading fluency as a test for new programmers. In relation to his own talents, Bill Gates stated:

> In 1975, I would have said, 'Hey, watch out, I can do anything.' I really thought I could, because I had read so much code, and I never found a piece of code that I couldn't read very quickly. I still think that one of the finest tests of programming ability is to hand the programmer about 30 pages of code and see how quickly he can read through and understand it. (p. 83)

Therefore the speed at which an individual could read and comprehend technical computer code was highly valued by this group.

An excellent memory and quick thinking are aspects that lend themselves to a person having good logical reasoning and analytic skills. When asked to talk about how they approached problems in programming, some were able to clearly state the logical structure they followed, like Charles Simonyi: "We follow these four steps of problem solving: First, understanding the problem, then devising a plan, carrying out the plan, and, finally, looking back" (p. 19).The steps he described are very similar to the steps espoused in Bransford and Stein's (1984) handbook, *The Ideal Problem Solver*. All of the participants talked about their step-by-step process they used while working on problems. At this point in time, the field of computer science was dominated by mathematicians. This background affected the analytic thought processes of this sample, depending on which person the interviewer spoke to. As Bill Gates observed:

> Most great programmers have some mathematical background, because it helps to have studied the purity of proving theorems, where you don't make soft statements, you only make precise statements…. Math relates very directly to programming, maybe more so in my mind than in other people's minds, because that's the angle that I came from. (p. 81)

However, not everyone was as math-oriented as Gates, and participant responses seemed to fall along a continuum of math and computer science way of thinking.

Some programmers used more colorful analogies to describe the way they liked to solve problems. C. Wayne Ratliff described his own interpretation of programming:

> I used to play a mental game and ask myself, what would I be if I'd been born a hundred years earlier? I don't know, but one possibility is a detective, because there's a lot of detective work in programming, particularly in debugging. You work with hints, clues. (p. 129)

Reminiscent of the fictional character Sherlock Holmes, Ratliff emphasized how

important deductive and inductive thinking were to working with computers.

Programming was the ultimate puzzle, and using good judgment and making the right

choices with a problem is another intellectual ability. Peter Roizen talked about this

distinction at length:

> I've always enjoyed games like chess and backgammon, because they require strategy
> to win, and because I always know at the end of the game whether I won or lost.
> Programming is very much like a game. If I just want to write a program to solve a
> problem without a lot of fuss, and there aren't many decisions to make, I can usually
> solve the problem in a fairly straightforward fashion. But there are always different
> directions to take. Programming is fun because it's challenging to make the best
> design decisions. (p. 192)

All of these intellectual abilities lend themselves to working with computer programs.

Creative.

Creativity can be separated into two different concepts: Creation of something

completely novel, or the recombination of old ideas into something new, (Feldman,

Csikszentmihalyi, & Gardner, 1994). At the beginnings of the computer revolution, both

types of creativity were exhibited as new programs and machines emerged. These

different facets of creative thinking manifested on a continuum with an artistic side and

an engineering side, and most participants in this sample displayed a combination of

both. Andy Hertzfeld found this balance exciting:

> It's the only job I can think of where I get to be both an engineer and an artist.
> There's an incredible, rigorous, technical element to it, which I like because you have
> to do very precise thinking. On the other hand, it has a wildly creative side where the
> boundaries of imagination are the only real limitation. The marriage of those two
> elements is what makes programming unique. You get to be both an artist and a
> scientist. (p. 260)

As important as intellectual abilities were to getting started in the computing field, having exceptional creative abilities helped separate the truly talented individuals from the average person.

Imagination was often mentioned as a key creative component. Ray Ozzie, "There is a tendency for people to write programs just to make money and not to solve problems. Instead of being innovative, they look at what already exists and try to copy it thinking, 'Gee, I can do a better one of these'" (p. 177). Charles Simonyi talked about the open-ended creative process he used to start off solving problems:

> The first step in programming is imagining. Just making it crystal clear in my mind what is going to happen. In this initial stage, I use paper and pencil. I just doodle, I don't write code. I might draw a few boxes or a few arrows, but it's just mostly doodles, because the real picture is in my mind. I like to imagine the structures that are being maintained, the structures that represent the reality I want to code. (p. 15)

Creative brainstorming about problems was how many new ideas were spawned. At this time, there was a push to try new things because there were still so many possibilities to explore. As Michael Hawley put it, "So many new things can be done with computers. I think often great new ideas come from recombining old ideas in new ways, and if there's one thing a computer's really good at, it's taking huge piles of information, mixing it together, and letting you play with the results" (p. 315). The digital environment was well suited for this playful aspect of creativity.

Inventiveness was another aspect of creative thinking brought up in the interviews with this group. There have always been tinkerers, but computers gave them a new way to explore the realm of the unknown. Ray Ozzie put it this way:

> Programming is the ultimate field for someone who likes to tinker. Tinkering requires tools. Electrical engineers have various components they can put together to build something. But they're constrained by the availability of physical equipment. With a computer, if you can think about it, you can do it. You can design your own tools or

create the parts as you go along. If you don't like something, you can just change it or rewrite it. (p. 188)

In the computing world, tinkering and inventing were synonymous, as Butler Lampson elaborated, "It's much, much harder to make things happen in other fields. If you're a physicist, you have to live with what nature has provided. But with computer science, you invent whatever you want. It's like mathematics except what you invent is tangible" (p. 26). That was why this time period was a crucible of new ideas; there was a freedom to be creative with whatever new tools could be invented. As Gary Kildall put it,

> A lot of programming is invention and engineering. It's much like a carpenter who has a mental picture of a cabinet he's trying to build. He has to wrestle with the design and construction to get it into a physical form. That's very much what I do in programming. (p. 65)

Many of the programmers in this sample could fall under the category of inventor because they created new products that went on to change society.

The final creative aspect emphasized by the programmers was having a sense of aesthetics in relation to the computer code. Recognizing beauty in the computer set the expert programmers apart from the novices. It was something that one could instantly recognize, like Charles Simonyi: "I think the lister gives the same sort of pleasure that you get from a clean home. You can just tell with a glance if things are messy – if garbage and unwashed dishes are lying about – or if things are really clean" (p. 13). Comparisons to artists popped up often, as when Ratliff described his own process of coding:

> I like to make an analogy between writing code and sculpting a clay figure. You start with a lump of clay and then you scrape away, add more clay, then scrape away again. And every now and then you decide that a leg doesn't look right, so you tear it off and put a new one on. There's a lot of interaction. (p. 120)

Bob Carr also saw himself as an artist, "I spent a lot of time thinking, scribbling – a lot of subconscious activity. That's where the art is. In art, you can't explicate how the end result was achieved. From an artistic standpoint, the best software comes from the realm of intuition" (p. 217). The artistic appreciation of code was an important aspect of technology talent that these programmers felt was essential.

Just as with the intellectual abilities, creative abilities should not be the only area of strength in a programmer. Over and over again, the participants in this sample emphasized the importance of a balance of abilities and skills. In Subotnik's (1993) interview with Joseph Bates, he offered his own opinion of what qualities were necessary for a person to be talented in computer programming,

> One is a sign of creativity. That person can take ideas, find variations, blend ideas across disciplines or topics…. Then the other important thing is that, at other times, they are able to be very analytical, careful, and clear in their thinking. They can take the ideas that they or somebody else generated and filter through them to what's really there. (p. 318)

The most important idea that emerged from this sample was that the intellectual and creative abilities had to work together to produce an individual who was talented in programming.

Socioaffective.

Besides intellectual and creative abilities, an individual's social abilities also play a part in his or her development. Gagné (2003) emphasized a person's ability to communicate with others and perceive social cues and actions as falling under this category. Lammers (1989) did not directly ask the programmers in this sample about their social interactions, but some of them offered examples that related to this concept.

71

Most of the programmers saw their programs as a way to communicate their thoughts and ideas to the general public, and they hoped to produce programs that the users actually wanted. Frankston stated, "When you communicate, whether it be in writing or programming, what you say has to be understood by your recipients. If you cannot explain a program to yourself, the chance of the computer getting it right is pretty small" (p. 157). Clarity was important in relation to programming as well as interacting with other people. Toru Iwatani put it eloquently when he said, "You must understand people's souls (*kokoro*) and be creative enough to imagine things that can't be thought or imagined by others. You must be compelled to do something a little bit different than the rest of the crowd and enjoy being different" (p. 269). This desire to understand others but not completely conform gave Iwatani a unique perspective. The extent to which an individual felt successful in communicating with others depended on how much importance they placed on social interactions.

In relation to communication, many of these programmers were self-described nerds. But Jaron Lanier warned the readers from thinking that all programmers could be described the same way: "It's very hard to generalize, in spite of the stereotypes. Like the one that programmers all dress terribly, stay up all night, and all that. There's a current generation that is pretty much oriented toward the Macintosh type of thing" (p. 300). A classic example of this stereotype was the inability to interact with the opposite sex. Andy Hertzfeld described the interview process he participated in once for a prospective new programmer:

> As soon as the guy walked into the room, I knew it was going to be problematic, because he seemed extremely straight-laced and uptight, dressing more like an insurance salesman than a technologist. He also seemed very nervous as he fumbled at our first few questions.

I could tell that Steve Jobs was losing patience when he started to roll his eyes at the candidate's responses. Steve began to grill him with some unconventional questions. "How old were you when you lost your virginity?", Steve asked.
The candidate wasn't sure if he heard correctly. "What did you say?"
Steve repeated the question, changing it slightly. "Are you a virgin?" Burrell Smith and I started to laugh, as the candidate became more disconcerted. He didn't know how to respond. (Hertzfeld, 1982)

Graham (2003) wrote an essay called *Why Nerds are Unpopular,* where he hypothesized that nerds had more important things to think about than dressing and acting in a way that would attract a girlfriend or boyfriend. Although a person's virginity was brought up jokingly by several programmers, this was not an aspect that most of them felt was necessary in someone with technology talent. More emphasis was placed on the amount of time a person had to devote to computer programming, and how that could take away from time that might be spent on romantic relationships.

For all the negative social stereotypes about computer programmers discussed, there was one aspect of socialization that the programmers enjoyed. They knew tricks with technology that the average person did not, and so they were happy to use their "magic powers" to show off their abilities. This word was mentioned by three programmers, and they talked about how much they enjoyed displaying the secrets behind the magical grey box. As Andy Hertzfeld put it, "More than anything, I loved impressing people. I could show my friends something and watch their mouths drop as they said, 'Oh, wow.' Graphics and sound were two things most people took an interest in. I guess I wanted to impress myself more than anyone" (p. 251). Because most of the men in this sample worked for computer companies, part of their ability to impress people came out in the products they created for the public.

The participants in this sample admitted that it was hard to escape the stereotype that all programmers had high intellectual abilities and low social skills. To combat this stereotype, they were forced to find ways to successfully communicate with people, and computers allowed them to do that. Michael Hawley said, "Communication is a beautiful thing. It's why telephones and personal computers were such an instant hit. The computer is probably the ultimate tool for looking at the problem of communication" (p. 313). Even if these individuals were still considered "nerds," they were able to overcome most of their social problems through the use of computers.

Intrapersonal Catalysts

   Motivation.

The category of motivation included a person's needs, values, interests and goals. What motivated these individuals to pursue technology depended on different factors in their lives. The programmers in this sample discussed their goals as successful businessmen as well as different internal and external motivators.

The goals described by those interviewed ranged from developing the perfect program to making the experience for the software user as smooth as possible. Ratliff explained: "There's a spectrum of programming. At one end is a programmer who's working 100% for the user; at the other end is a programmer who's working on some mathematical problem and couldn't care less about the user" (p. 122). There was a varying degree of responses that fell closer to one side or the other. In some ways their goals could be very broad, like with Raskin, "When I created Information Appliance, I was no longer trying to make computers. I just wanted to make the benefits of computer technology easily available to everyone" (p. 235). It was the goal of other talented programmers to reduce the frustration of the average user as much as possible. Page explained, "Flying is like programming. A good pilot gives the passenger a flawless, boring ride, while a good programmer gives the customer a flawless, boring experience on their computer" (p. 106). By 'boring' Page meant that the user did not have to think about the inner mechanics of the system because everything just worked as it should.

Although they wanted their products to be successful financially, it was the internal motivation that sustained them and kept them enthralled with computing. They described specific moments in their lives related to computing that they considered to be peak experiences. For Jonathan Sachs, it was a defining moment at the very beginning of his career:

> That was when I really got interested in programming. It was my first programming job where I was able to sit down and actually have a conversation with the computer. To have a computer respond the way you expect it to is exciting. There was a certain kind of electricity about that experience. I remember it very clearly. (p. 164)

Andy Hertzfeld also mentioned the exhilaration of his early experiences, "I found I had a talent for programming. A computer gives an amazing feeling of control and power to a kid. To think of something, and then get the computer to do what you thought of, was such a great feeling. It always has been. That's what attracted me to the field" (p. 250). The situations described in these quotes fit into the definition of Maslow's (1998) concept of peak experience where, "The person in the peak experience usually feels himself to be at the peak of his powers, using all his capacities at the best and fullest" (p. 117).

This feeling of exhilaration was echoed by other programmers, such as John Warnock: "It's just great fun; it's very satisfying, sort of like mountain climbing. It's like a lot of activities in life: When you're successful at it and you finally get it to work right, it feels good and you get a big kick out of it" (p. 49). This thrill made them feel powerful, but John Page had a warning for young programmers, "Some people get such a high from completing the project, they roll straight to another one. Doing that will literally burn you out. Each time you go straight to a new project, you get worse at it" (p. 107). As long as computer software businesses acknowledged the power of these emotions, they could take steps to ensure the mental balance of all of their programmers.

Volition.

The individuals in this sample expended great effort to start their businesses, and sometimes success or failure rested solely on their shoulders. It was an expectation in this sample that all of their efforts were to be placed into computer programming, often at the expense of their families. John Page said:

> You have to say to all your relatives, 'Look, I'm going to be gone from six to nine months. I'll be here physically, but I might as well not be. I'm going to be working on this thing and I'll be absent-minded. I want you to understand and tolerate that and I promise I'll make it up to you when I get to the other end.' If you have loved ones it's important to come through on that promise. Working so hard can be devastating to your marriage and to other relationships. (p. 107)

The concept of normal work hours was almost foreign to them, and it seemed that the best progress was made in the wee hours of the day. Scott Kim stated, "You keep all these thoughts and algorithms in your head and you're off in a corner trying to sort through them and put them together in a giant puzzle, working long nights" (p. 278). The individuals in this sample were definitely not slackers and placed high value on responsibility and hard work, a message they wanted to pass on to the next generation. Bill Gates observed that:

> Programming takes an incredible amount of energy, so most programmers are fairly young. And that can be a problem, because programming requires so much discipline. When you're young, your goals aren't as stable; you may get distracted by one thing

76

or another. Young programmers should stick with it, though, and they'll get better. (p. 83)

This perspective was from someone who was able to achieve being a successful businessman as well as an extraordinary programmer.

Another quality related to volition is persistence. This meant that a person would not be deterred from working, even when faced with repeated errors that came up time after time. It was accepted that problems would occur while working on computers, and it was the programmer's job to 'debug' them. C. Wayne Ratliff enjoyed this process of error correction:

> The moment of programming I enjoy the very most is when I get something almost complete. I try it for the first time, it fails miserably, and it continues to fail until about the one hundredth time, when it does pretty good. There's a peak experience there, because then I know I've got it. I just have to apply a little more elbow grease to weed out the rest of the bugs. (p. 122)

Jonathan Sachs emphasized persistence as the aspect he valued most in a good computer programmer:

> It's a combination of talent, temperament, motivation, and hard work. I find quite a lot of people expect to be really good after a short time, but I haven't known too many people who have been successful in doing that. Success comes from doing the same thing over and over again; each time you learn a little bit and you do it a little better the next time. (p. 170)

Hard work and persistence were qualities that this sample thought helped a person find success and fulfillment in the world of computing.

Self-Management.

The qualities of initiative, autonomy, concentration, and efficiency, all fell under the heading of self-management. 11 out of 20 programmers specifically mentioned being self-taught when asked about how they initially learned about computer technology.

Relying on one's own interest in pursuing knowledge was important to these programmers, like Toru Iwatani: "I had no special training at all; I am completely self-taught. I don't fit the mold of a visual arts designer or a graphic designer" (p. 265). This desire to explore and learn more about the world led these programmers to learn on their own and create the technology that they wanted. Peter Roizen, clearly stated his opinion about formal versus informal learning:

> I'm not a big believer that studying computer science makes good programmers… maybe you're better off learning that information on your own, like I did. Well, I had to teach myself, because there weren't any reference books or classes at the time. I'm saying this because most of the problems that occur don't fall into the categories you learn in a book or a class. (p. 202)

Roizen implied that he had to learn on his own by default because there were no formal resources available to him at the time. But others actively sought out anything they could get their hand on, like Bill Gates, "The best way to prepare is to write programs, and to study great programs that other people have written. In my case, I went to the garbage cans at the Computer Science Center and I fished out listings of their operating system" (p. 83). Instead of passively sitting there and being instructed by a teacher how to program, Gates took the initiative for himself.

Concentration and efficiency were two other aspects that were brought up during these interviews. The ability to stay focused and not waste time were qualities greatly valued by these programmers. Andy Hertzfeld emphasized how important he thought these qualities were to his work:

> To do the kind of programming I do takes incredible concentration and focus. Just keeping all the different connections in your brain at once is a skill that people lose as they get older. Concentration is a gift of youth. As you age, you become wiser. You get more experienced. You become better at living. But I don't think that when I get older I'll have the edge I do now. (p. 260)

Peter Roizen also talked about focusing on problems and not getting distracted: "Being practical is important to good programming. You must be able to guess accurately how long it will take to complete a project, have the ability to calculate what can and cannot be done in that time, and then do it, resisting the temptation to go off and do other things" (p. 193). Initiative, autonomy, concentration and responsibility helped to make the individuals in this sample successful in their businesses.

Personality.

In Gagné's (2003) model, he emphasized sensitivity, adaptability, self-awareness, temperament and individual traits as being part of one's personality. I also included the aspect of being a non-conformist under this heading, as this was a quality emphasized by close to half of those interviewed. Dan Bricklin reflected, "I was usually working from 11:00 in the morning until 1:00 at night. I wore torn blue jeans, beard out to here, and hair way down my back." He took pride in living outside of the norms of society.

In the 1980s, programmers flocked to California to become part of the new businesses there, and they all seemed to have similar personality traits: "Silicon Valley entrepreneurs typically exhibit unique traits and qualities not found in other areas of the country. They scorn tradition, disregard personal self-preservation, dismiss nay-saying authorities, take inordinate risks, intuit new concepts, and persevere beyond the ordinary" (Landrum, 1993, p. 22). Lammers (1989) asked about these non-conformist personality types that seemed to be heavily represented in this sample: "There's a lot of talk about how large software companies find it difficult to attract talented people who can produce great software, because those mavericks are so independent that they want to work on their own" (p. 75). These strong-willed personalities could be a problem in a collaborative environment. However, these individuals worked to balance everyone in a way that best suited their working style. This sentiment was found in the interview with Ray Ozzie, who spoke from experience:

> Many managers find programmers difficult to work with. I've seldom had a problem. Problems most often seem to occur when management tries to issue edicts or excessive controls. Programmers are very creative, self-directing, self-motivating people. You have to recognize that up front, and resign yourself to getting in their way only when it's very important to do so. (p. 178)

When the majority of technology workers in a business have a certain style, they tend to create the culture around them to suit their personalities. In the book *Winning the Technology Talent War*, Brantley and Coleman (2001) concluded that, "The most important reason why a tech worker stays with a company is because of the culture, and feeling like the work is challenging is second" (p. 5). It was beneficial to the managers of these software companies to create an environment where these different personalities felt comfortable.

Besides the standard range of personality traits, this sample seemed to have a lot of quirky traits that set them apart from other people. This slightly *oddball* way of being can be puzzling to people in normal society, but can actually be positive for creative inventors. In a book dedicated entirely to quirky geniuses, Suran (1978) described some of their characteristics: "Oddballs are comfortable with a unique experience of reality, and they may even take pleasure in inflicting this separate sense of reality on others" (p. 214). Instead of shying away from their different way of thinking, they reveled in it and sought to make others aware of their unique perspectives. For a programmer, being different did not always have to manifest itself as the quiet nerd stereotype. Ray Ozzie advised the next generation of programmers to, "Don't worry if people think you're weird" (p. 189). The types of personalities and temperaments present in this sample were varied, but their passion for computer programming brought them together. Sometimes they were very similar and other times they were able to work in a complimentary fashion.

Environmental Catalysts

   Family.

Many of the interviews done by Lammers (1989) focused on the current experiences of the programmers, but several of them discussed how their family influenced them at an early age. Because there were not as many computing opportunities in the first half of the twentieth century, these people had to get inspiration from tinkering with what they could find around the house. Even if it was not necessarily related to engineering, they sought out any opportunity to create. Gary Kildall, recognized the importance of these experiences in his own childhood:

> Seymour Papert has this notion that kids learn to be inventive by tinkering with gears and other mechanical gadgets. The skills you learn and practice with this kind of play carry over into other areas. Papert is certainly talking about my childhood experience. My father was a great craftsman. I used to stay and watch him by the hour, and then I would go outside and try to imitate him with my own hammer and nails. (p. 62)

Although family influences were talked about sparingly, if someone had been profoundly influenced by their parents, they did not hesitate to mention it. Charles Simonyi related how he used his father's connections at work to get as close to computers as possible: "My father was a professor of electrical engineering and this engineer was a student of my father's. I think my father asked him to let me in once as a favor. I made myself useful. First, I brought him lunch, then I held the probes, and finally I offered to be a night watchman" (p.10). Simonyi was one of the few in this sample who had access to a mainframe computer at a very young age and whose father was in the same field in which he was interested.

In another case, it was a different family member who helped the programmer get into computing. After seeing his interest in physics and electronics at an early age, Joseph Bates got an opportunity to advance even further: "When I was eleven, I got introduced to computing through a cousin of mine who worked in a university laboratory, and I really liked that" (p. 314). The universities seemed to be the door to the world of computer science, and Bates was lucky enough to know a gate keeper. Overall in this sample, the family influence generally manifested itself as parents who were supportive of their child's exploration and curiosity.

School.

There was almost no mention of high school as an influence in the computer technology talent development of this sample. At the time these participants were in school, computers were not a part of standard public education. However, the influence of a mathematics class was mentioned by one participant. Ray Ozzie recalled one particular teacher who encouraged his interest in technology:

> I started my freshman year of high school in 1969. One of the math teachers brought a programmable calculator to the classroom to show us, thinking we would get excited about this neat little toy. He invited us to see him if we wanted to play with it. A couple of us did, and we spent a lot of time exploring the capabilities of the calculator. (p. 179)

Ozzie brought up this memory because it was particularly salient to him and stood out as a memory of a teacher who helped him in this subject. Lammers (1989) did not ask a specific question about the high school experiences of these programmers, and therefore data related to that area was not collected.

82

There were very few personal computers at the time when these people were

growing up, and so many programmers had to wait until college to start learning about

computer science. Dan Bricklin felt computer training was important for programmers:

> People who have formal training often have an advantage over people who don't.
> Knowing what people have worked out academically is often very helpful. You
> know, some people have green thumbs and some people don't, but it helps if you
> learn a bit about the area first. (p. 150)

However, John Warnock came out with a dissenting opinion against formal computer

science training:

> My saving grace in life is that I was not introduced to computers at an early age. ... I
> went through university, all the way to the master's level, so I got a good, solid liberal
> education. I believe it's really important to have a very solid foundation in
> mathematics, English, and the basic sciences. Then, when you become a graduate
> student, it's okay to learn as much as you can about computers. (p. 49)

He believed that if a person just studied computers, then that individual would not have

any new insights to bring to the field. Having too narrow a focus could turn into tunnel

vision, and Warnock preferred to work with people who had more diverse and creative

experiences.

Because all easily accessible computers were owned by universities, they became

gathering places for the most avid programmers. Ray Ozzie, who had also been

introduced to computer science at college, stated, "Every night at 10:00 PM the machine

became available, and a small clique of people like me would work all night until our

time was up at 6:00 AM. We did this for years. It was great fun" (p. 179). But much more

than a kinship was formed at these gatherings, and each person would share their ideas

and discoveries to help others understand the possibilities held within the machines. Scott

Kim talked about how much knowledge he gained from his time at Stanford:

The sequence of computer-music courses gave me access to this wonderful community of people who were just out there hacking their hearts out for the pure love of it, and who would, at the drop of a hat, just spend the whole afternoon telling you what they were doing. That was a wonderful way to learn. (p. 275)

The programmers in this sample felt that the universities were where the most interesting learning was going on, both formal and informal.

Peers.

Peer influence was cited by many programmers as being a major factor in opening their eyes to the world of computing. Some of them had computing peers in high school, but the majority of them only found people like themselves once they went to college. For some, this was their first experience with computers, like Gary Kildall: "I originally planned to be a high school math teacher, and started taking math courses at the University of Washington. But a friend of mine had this FORTRAN statement card, showed it to me, and told me it was going to be a really big thing. I became so intrigued I had to get into it" (p. 60). The air of excitement that surrounded computer technology motivated people to share it with their friends. Once that passion had been ignited there was no turning back, as was the case for Andy Hertzfeld:

Like a lot of people, I hadn't even heard about computers until a friend told me they existed. He showed me some listings but even then I didn't have a clear idea of what a computer actually was until I took a class the following year. Then it clicked in my mind and I fell in love with them. (p. 250)

People wanted to share their excitement with others in their social circle. This collaborative excitement carried over into their first ventures in business and fueled the competitive race to create the best personal computer.

The new software companies like Microsoft and Xerox PARC became a magnet for the best and brightest technology workers. Bill Gates agreed that like minded people

84

want to work together, "I think most great programmers like to be around other great programmers. When they think up an incredible algorithm, they like having peers who can appreciate the cleverness that went into it" (p. 75). Because the presidents of these companies were often programmers themselves, they knew what to look for when hiring employees. With programming, it sometimes came down to a person's reputation among their peers, as emphasized by John Warnock:

> It's like hiring a writer. You can't tell how well someone writes simply from an interview. But if the person has written a number of prize-winning novels, you have a much better understanding of their talent. So we hire most of our people through the grapevine. When you've been in the business a long time, you get to know a lot of the people. You can pretty much find who you're looking for by their reputation in the industry. (p. 48)

In the software industry, the programmers had to learn how to work as a team. It became clear that one person could not do it all and that tasks needed to be delegated in a way that made sense to the different types of workers. John Page laid it out clearly when he described the two types of thinkers he needed in his company:

> Companies need people who are true architects – people who can see beyond the basic tools to understand what should be built, who understand which technical things to take advantage of and how…. Companies also need people who can write lumps of code for the grand architecture – people who are just good mechanics…. The question then becomes how to get the right balance of these two kinds of people. (p. 98)

The emphasis here is that these talented peers had to work together and compliment each other's talents to succeed in the world of computing.

In certain situations, a dyad ended up forming with two great minds working together in a complimentary fashion. Bill Gates related a particular partnership he had that he found to be most rewarding:

> There's been a mixing of ideas between Paul Allen and me, because so much of the programming we've done, we've done together. It is nice to have somebody who's up to speed to talk to when you're debugging code or you aren't sure about some

particular trade-off. In a sense, it's a way of taking a break, relieving the intensity without having to switch topics, just going in and discussing it with somebody. In the creative process, it's good to have the pressure off a little bit and yet still have your attention focused. Paul and I learned how to work together in an effective way. You don't find too many partnerships like that. (p. 84)

That partnership dynamic could provide healthy competition as well as a different perspective when it came to working on problems. Whether it was a best friend who shared their enthusiasm or a dedicated team of technology workers, these programmers relied heavily on their peers for the success of their companies.

Media.

Comments about the effect of media on their development were mentioned sparingly by the programmers. It was Andy Hertzfeld who described at length how the image of computer science changed for him:

In 1975, a programmer was someone who worked for a bank and maybe NASA, something a little more glamorous, but the career was really obscure. There was no reason your average teenager would want to be a programmer. The concept of the computer had not reached the average person, except in a science fiction sort of way. In 1975, no person could afford to buy a computer. In fact, the personal-computer revolution was just beginning then. That fall the Altair kit was put on the market, and I began to see articles in magazines about microcomputers. That development was very important to me. (p. 251)

Before that time, there was no excitement about computer technology. But once magazines and other media outlets began printing stories about the new types of computers, the atmosphere changed. Having a certain 'buzz' about computers peaked the interest of several of the participants. For Bill Gates, it was the media that spurred him into completing his first software program: "Paul Allen had brought me the magazine with the Altair on it, and we thought, 'Geez, we'd better get going, because we know these machines are going to be popular.' And that's when I stopped going to classes and we just worked around the clock" (p. 79). Once he saw the future of personal computers, he foresaw how important programs like his would be to this new technology. At the beginning of their careers, computers were only featured in technical magazines, but by 1982 they were featured on the cover of *Time* magazine as "Machine of the Year." Not only did media affect them, but the subjects of this sample were the ones making the headlines.

Talent

The combination of intellectual abilities, intrapersonal qualities, and external influences all played a part in the development of technology talent for these individuals. Gagné (2003) posited that only the top 10% of people who started off with certain gifts were able to transform those abilities into talents. The interviews in this sample were from the book titled *Programmers at Work* (1989), therefore the participants mostly talked about their talents in the context of their work environment.

It was clear that business success was at the forefront of many of these programmers' minds. They wanted only the most talented people working for them at their companies. Jaron Lanier explained his perspective: "Computers don't have any quality in themselves. They're absolutely empty things, tabula rasa. Since they are such empty minds, it depends entirely on the person involved, more so than in any other field or human endeavor" (p. 295). The software programs they created were only as good as the people who worked on them. Ray Ozzie stated,

> I come to work because I enjoy playing with computers, but I design the products as I do because I feel that I can provide the consumer with something useful. It is extremely important to remember that end goal throughout the product-development cycle. (p. 187)

If the users were unhappy with their computer programs, then they were unlikely to buy from that business again.

For some programmers, they expressed their talents by wanting to achieve perfection and create tangible solutions to difficult puzzles. As Raskin said, "With the Mac, I was trying to make the best computer I knew how to make" (p. 235). It was hard to achieve perfection when programmers would constantly find new puzzles and new paths to follow with a program. At some point, they had to make a final decision on when a product could be released to the public. A programmer could get caught up in trying to make something too perfect, as Roizen said from experience:

> Practicality is important when you're trying to make every piece of work the best because you always have a project when you're consumed with making it the ultimate in elegance. But if you're still working on it, it doesn't sell too well. You don't have a product, and that makes it difficult. (p. 193)

It was this search for perfection that kept these talented programmers going, and they constantly found new ways to use their knowledge and skills. Some of the software companies eventually folded, while others like Microsoft became extremely successful and have continued to dominate the computing world almost thirty years later.

Predictions for the Future

Back in 1989, these programmers already knew that computers had made an indelible impact on society. Almost all of them had a positive perspective on the future of technology and its impact on the next generation. Jaron Lanier stated:

> We are in a transition period. Until now, what we have wanted from life had to be got by manipulating physical matter. Now, we are just starting to organize our lives according to information. Eventually, our very experiences will be generated by information instead of the other way around. That will be the true information age. (p. 299)

The internet and World Wide Web were not publicly available yet, but already the people in this sample knew that the Information Age had begun to take over from the Industrial

Age. Butler Lampson recognized that changes in society take time, and that the biggest

transitions had yet to take place:

> I give a talk to general audiences titled: *The computer revolution hasn't happened yet.*
> Its basic theme is that major changes in the world take a long time. … I think this is
> true for the computer revolution. People like to think that we're moving much faster
> because we go from idea to finished product in six months. That's nonsense.
> Computing is just beginning to become a significant part of the economy and just
> starting to affect people's lives. (p. 36)

Lanier continued with his predictions about the way people would communicate with

each other: "People should be able to speak and breathe programs just like they talk now.

Making little worlds inside the computer should be as easy as saying hello to your friends

in the morning" (p. 289). The way in which society has become connected with mobile

devices and cell phones reflect this prediction.

Another development in the field discussed by these programmers was the

concept of universal computer literacy. John Warnock said, "Computer literacy will have

to become essential. People deal with computers every day and don't realize it" (p. 50).

He could see that computer use was not limited to programmers anymore, and every

person in the country could someday derive benefits from this new technology. Butler

Lampson had a different opinion:

> To hell with computer literacy. It's absolutely ridiculous. Study mathematics. Learn
> to think. Read. Write. These things are of more enduring value. Learn how to prove
> theorems: A lot of evidence has accumulated over the centuries to suggest this skill is
> transferable to many other things. To study only BASIC programming is absurd. (p.
> 38)

He expressed a fear there that other areas of knowledge would deteriorate if computing

became the sole focus of society. Most programmers remained optimistic about the

positive effect of computers on future generations. Bill Gates specifically had high hopes

for the field of computer science:

Really brilliant people are getting involved and contributing: programming is much more of a mainstream activity now. The fact that people are getting exposed to computers at such young ages now will help change the thinking in the field. A lot of great programmers programmed when they were in their teens, when the way you think about things is perhaps most flexible. (p. 81)

Ray Ozzie was also thinking of the new generation of programmers and how there was going to be more access to technology than ever before:

I'm enthusiastic about personal computers because younger people with programming in their blood won't have to fight and play politics to get machine access as we did. I hope parents and schools recognize what a wonderful resource PCs are. I would love to see computer centers made available for kids who want to play but can't afford machines of their own. (p. 183)

Finally, Jaron Lanier had a vision of a new era, and was very hopeful for the future of technology:

In the next few years, life is really going to change. It's very exciting to think that it's the young kids, the generation being born right now, who are going to grow up with this new technology. They are going to be the ones to really benefit from what we've been talking about. (p. 300)

The generation of children he referred to grew into the teenagers who the researcher interviewed for the contemporary sample in this study.

Summary

All of the people interviewed in this sample placed high emphasis on intellectual and creative abilities, especially when it came to problem solving and software creation. However it was the intrapersonal qualities of these individuals that distinguished them as having CTT above and beyond the average person. The programmers who were successful were extremely driven and goal-oriented, and put all of their time and effort into developing new computers and technology. The ambition and motivation shown by most of this sample are qualities that also relate to *entrepreneurial giftedness*, a concept elaborated on in Chapter 5 (Shavinina, in press).

The participants had a fairly universal perspective on the qualities required for computer technology talent. Bill Gates offered this insight into the development he saw in people like himself:

> It's amazing to see how much great programmers have in common in the way they developed, how they got their feedback, and how they developed such a refined sense of discipline about what's sloppy and what's not sloppy. When you get those people to look at it a certain piece of code, you get a very, very common reaction. (p. 83)

The only external influence that these participants cited as making a significant impact on their lives was working with their computing peers. The individuals with brilliant minds and a desire to learn more about technology came together at universities and computer companies and shared their ideas. They had high hopes for the future of technology, and they had the power to make their computer creations a reality.

Sample 2: Snapshot

The participants in this sample (N=9, Male=9) were individuals with current

technology jobs or who had technology jobs in the recent past. They volunteered to

participate in this project after being contacted by the researcher through different on-line

networks. The quotes presented below were taken from e-mail responses, on-line chat

transcripts, and audio taped conversations to the interview questions in (Appendix A).

Their age at time of interview, location, notable achievements, and whether or not they

received any gifted services are presented in Table 2.

Information Table for Sample 2

Name	Achievements	Gifted	Age	Location
"Escher"	Aerospace engineer and programmer, works for NASA	Yes	28	Lawrence, KS
"Felix"	Helped design an IP backbone for a fiber optics company	Yes	30	Washington, DC
"Hugo"	Software programmer for an automotive company	No	31	Memphis, TN
"Slate"	Web applications developer, web designer	No	27	Denver, CO
"Pike"	Left a technical support job and recently returned to college	Yes	31	Youngstown, OH
"Argus"	Professional software engineer, accomplished digital musician	Yes	34	Seattle, WA
"Oberon"	"Computer guru" for 20 years, currently working as a butcher	Yes	28	Toronto, Canada
"Sako"	Computer tech for a software / networking / hardware company	Yes	29	Springfield, MO
"Moby"	Studying education and computational linguistics, has MFA in classical piano	Yes	28	Tokyo, Japan

Natural Abilities

Intellectual.

The participants interviewed in this sample brought up intelligence, memory, logical thinking and analytic problem solving as important to working with technology. The interviewer asked if they had been involved in gifted classes when younger to get an initial gauge of their intellectual abilities. Seven out of nine were in some sort of gifted program, and the two remaining participants did not have access to services or were not in gifted classes by choice.

According to this sample, a person needed to have a variety of intellectual abilities to be a good *techie*. Pike's comment is typical of this group: "Hand / eye coordination, the ability to fix things. In general, good spatial / mathematical intelligence." Slate suggested that a simple test of technology knowledge could indicate the most capable person: "It's probably pretty easy to figure out who is technical with a simple quiz, much like the *Maximum PC* geek trivia published yearly." Knowing a large amount of information about technology was considered helpful, but Escher put much greater weight on having a high memory capacity:

> I'm not as good at the code as others, writing purely abstract code, database management stuff. I really need a visual representation to help, but you know, Chuck is one of those people he can see 10,000 lines of code in his head, I don't know how it works, but way before it's actually there, he sees it, how it's going to work. It helps in math too, and math helps that.

The friend he mentioned was in the gifted program with him and went on to start his own software company. However, this level of memory was considered very unusual, and Escher was the only one in this sample to mention it.

There were other individuals who spoke clearly about the logical steps they followed, using inductive and deductive reasoning to troubleshoot technical problems. Felix outlined his problem solving technique:

> I can usually take a top-down approach to finding the source of a problem and figuring out what the best solution to it will be. My rule is to start with the basic stuff first (is the machine plugged in?), then move on to higher levels of operation. Until something sticks out as being obviously wrong, then I change that thing until stuff starts working.

An emphasis was placed on logical thinking, which could manifest in a mathematical format for some. Moby even used an equation to describe his problem solving:

> Technology, like any f(x) equation, is simple. Behind it all is what you want it to do, the output. Input is what is arbitrarily necessitated based on the physical means and limitations and advancement that exists during the period. If I have a good idea of what I'm inputting and what is on the input medium, I infer a great deal from what the function is doing.

When asked what skills make a good computer programmer, Argus used an analogy to compare talent and training:

> There are the typical answers like an analytical mind or a penchant for solving puzzles. And then there are the simple answers like you have to be intelligent. To me, though, I guess if I were to have to say anything, it's talent. Anyone can learn how to be a 'techie' and they can do it well. But the difference between the people with talent and the people with mere training is the same difference between Google and Bonzi Buddy.[3]

But Felix also emphasized that understanding the bigger picture was more important in the long term: "Learn as much as you can about how and why things work the way they do as opposed to just learning how to make machines do things. The latter will eventually run you up against a wall that will be insurmountable at a certain point without the information gleaned from the former." Overall, the intellectual abilities mentioned by the

---

[3] Google is the most popular search engine used worldwide, and Bonzi Buddy was a computer application that made it onto *PC World*'s top 20 most annoying tech products (Tynan, 2007).

individuals in this sample formed the base level from which a person could start developing technology talent.

Creative.

The aspect of creative problem solving was emphasized by this sample, as well as imagination and a sense of aesthetics. Escher said that creativity was what distinguished human from computers: "There's things that machines are good at solving, but there's a lot of creativity to problem solving that I don't see machines doing for quite a while. And I think that's part of it, knowing what to apply when." The ability to generate ideas and solve problems with technology requires creative thinking, and Moby saw that as essential for technology talent, "Those students who have a mind for what they want to create and have an idea of how it can be done with computers, the sense of how to use and manipulate technology naturally follows."

Another specific aspect of the creative domain is imagination. Again Moby was emphatic about the importance of this aspect:

> In essence, I think that's really where true innovation and creativity happens – in the imagination. This always seemed to be one area that was really lacking in public education, the effort to spark students' imaginations. I think the trick is not to teach too much, but just enough that a students' desire to learn will take over and lead them to new areas that the teacher cannot follow.

This creativity can be a desire that fuels these talented students, as Sako wrote:

> These creators find the near limitless nature of technology to be an excellent environment for their creativity to explore and construct within. These are often the best software architects, the best in computer animation, the best in web development, etc.

Creativity as a cognitive quality gives a person the extra ability to imagine new possibilities with technology and make positive contributions to the field.

In a more artistic sense of the concept of creativity, four out of nine were involved in either art or music, some more deeply than others. Sako mentioned his anecdotal observation that, "The students who are gifted in technology often show a gift for music and languages. This creativity compliments a very curious personality that finds expression in the mathematical, especially in pattern recognition." Even if they were not directly involved with art or music, there was still an aesthetic nature to working with technology. One of the most original comments related to creative thinking came from Escher as he described an experience he had that was almost a hallucination:

> I don't know if you've ever looked at a brick building and said, oh it's made out of very tiny calculus kind of parts! I promise after getting out of some math exams as an undergrad, I'd walk out on the street and like… I don't know, I think the movie *The Matrix* captured it well, when he looks down the hall and sees the falling characters, you know. It's like, that's what I felt like sometimes. Is this real? Or am I actually… Am I seeing like the real behind what I normally think is an image, or is it vice versa?

He viewed the world around him through his own lens, and saw math and technology everywhere. It was creative thinking like this that these participants considered important to working with computers.

Socioaffective.

Most of those interviewed had high verbal skills and were able to clearly communicate their ideas to me in on virtual environment. However, it was hard to get a complete sense of their social abilities because of the limits of the e-mail medium (Fontana & Frey, 2003). Many did not talk specifically about their own social or emotional levels, but instead chose to focus on describing the stereotypical techie. Escher talked about his ability to communicate with other people, but was quick to point out that he was different from other computer programmers:

96

I never felt like I had a communications problem like a lot of other computer geeks, but I sure know those people. I mean, I'm sure I'm probably on average I'm shy and introverted and that kind of thing, but as far as like, the computer guy on *Saturday Night Live*, I know who that is and I don't think I'm that person. It's probably... a lot to be gained in getting technology kids to understand, well not understand, but at least practice talking to people that don't know what the heck they're doing in their own little microcosm, in their own little world.

It seemed that limited social aptitude was judged by this group as a defining characteristic of a techie, but they stopped short of applying that description to themselves. Only Moby was open enough to admit to his weak social abilities: "Socially I was a failure. Really the only place I could go was back home, playing video games and using the computer.... It just provided me a place to run away from then. But no man is an island, and we do need true human interaction."

The idea that computer nerds were unable to attract members of the opposite sex was brought up again during the interviews with this sample. It got to the point of parody when Argus responded to the question:

Unfortunately, the only way one can really tell if someone has a true 'technology talent' is to somehow find out if they're virgins by the time they're eligible for the draft. Or they're card-carrying members of the USCF. Or they can speak Klingon fluently.[4]

Csikszentmihalyi, Rathunde, and Whalen (1993) wrote about how this "virgin" perception appeared more frequently in gifted individuals: "Talented teens are more conservative in their sexual attitudes and aware of the conflict between productive work and peer relations" (p. 246). It seemed like it was almost impossible to have both technology talent and romantic relationships at the same time, as Moby joked: "I remember one friend who, as soon as he thought he had a girlfriend, turned his back on computers. Then when he found out she wasn't reciprocating, or just didn't like him, he

---

[4] Males are eligible for the draft at the age of 18 in the United States, the USCF is the United States Chess Federation, and Klingon is a language spoken by an alien race on the science fiction TV series, *Star Trek*.

came right back." These individuals would have liked to have girlfriends, but they just felt that computer programming was not an activity that attracted women. This issue of gender and technology is discussed further in Chapter 5.

In the interviews with this sample, the idea was brought up again that computer programmers could use their "magic" to impress people. Their knowledge about how to do amazing things with technology was one way to get positive social interaction. Escher explained it this way:

> I think there's a pretty common thread that I'm impressed by things just, how they work, why they work, not necessarily can be used for anything, but kind of the 'gee whiz'... There's always been kind of a magic. I always like being the magician, it's like you know the tricks that make something happen that people don't understand how it happens. That's got some allure to it, and I think that's pretty common among computer people, people that do well with technology.

Another participant related a story where the concept of having magic powers was also mentioned. Oberon went to a person's house to help network some computers: "I said 'Let's go fix your computer problem.' And we went inside, and I did. And he said 'Wow. You're like a wizard.' And I said, 'Yes, I am a Wizard.'" The feelings of excitement that come from impressing people are similar to stories mentioned in the historical sample. The quest to get that positive feedback from other people was a big social motivator for some of these participants.

The participants in this sample placed emphasis on high analytic abilities, creative problem solving, and imagination as qualities important to technology talent. They also remarked on the low level of social skills as common among computer programmers. Many of them talked about the nerd stereotype, and emphasized their positive natural abilities while downplaying their own negative qualities.

Intrapersonal Catalysts

Motivation.

When it came to the goals of this group, there was a variety of responses. In the area of achievement, some were motivated to learn as much as possible. Sako described his college interests: "I was a computer science / computer information systems / computer mediated communication / history / economics / sociology / philosophy minor in college. Somehow being a Renaissance person isn't looked upon with the favor of society the way it once might have been." However, the majority of respondents in this sample fell on the other side of the continuum, like Argus: "Once junior high came around and I had a choice of taking honors classes or regular, I went with regular every time. I didn't care about challenging classes; I was just fine taking classes that required me to do as little work as possible." Getting good grades in school was not an important goal to most of the participants in this sample.

The motivation to develop talent in technology could be internal or external. The participants mentioned passion and desire as a quality they felt was important. Sako commented, "They are driven much like Bach, or Monet, or Leonardo. They tend to be drawn to mathematical expressions of the world, so Bach and M.C. Escher are favorites." Later on Sako elaborated on this internal drive to explore, "Find your passion. Technology allows us to explore endless boundaries and scratch our own personal itches. At the deepest level it is transformative, much like finding religion." This sample had experience in many different tech jobs, and they knew what motivation was important to enjoy the work. Slate advised, "Don't do it for the money, do it because it's interesting to you. You do NOT want to be stuck in a data entry position!" Sako believed that an entire group of technologists were motivated by their desire to socially connect with other people in the world:

> This group is more driven by the fusion of technology with culture. The iconic representation of this type is Apple CEO, Steve Jobs. This group tends to be more focused towards technology as a social enabler and equalizer. They usually come from humanities backgrounds or linguistics. They are the builders of the Web, the core of community, and the defenders of technology culture and history.

Although the process of using computers could produce intrinsic rewards, there were also extrinsic motivators. Escher talked about working in his school computer lab with friends to fix the network so they could play games: "We basically had to troubleshoot it, figure it out, and at [high school], that was the extracurricular thing. We'd sit in there and spend hours troubleshooting it so we could play for 20 minutes." Oberon was even more direct about his motivation when it came to hacking: "Games. Then, later, porn. And a little bit of the anarchic spirit. That was the driving factor. Learning to program was about learning to crack games to play for free, learning to crack systems to get porn for free." There was the internal motivation to be rebellious, but it was also rewarding to see his efforts come to fruition. If they could outsmart the system and get something for it, then they would, as Oberon stated, "In high school, I hacked pop machines for free soda, phones for free phone calls, and the power company to shut down the whole town's electricity for 2 hours in the middle of a school dance." Many goals were a mixture of personal gain and the challenge of overcoming restrictions.

Volition.

Another important aspect that the interviewees mentioned was persistence and being able to work through mistakes without giving up. Felix described his trial and error approach:

I generally have to get my hands on something, break it if it is not already broken, and attempt to fix it. After a few iterations of break/repair I find I usually know enough about a thing to show others their way around it. Oh and if/when all else fails I make a point to find someone more familiar to assist with the repair, which again falls under the heading of 'know where to get the information you lack.'

He was not afraid to break technology in order to find out how it worked, because he was confident he could figure out how to fix it. Escher talked about how he saw a lack of persistence in today's schools and it concerned him:

There's a very complicated system there. I think that, if a kid tries something and they fail, or it doesn't work out the way they wanted it to, then they've got the sour grapes kind of attitude, and then that inspires the peer pressure that everybody has the sour grapes about it.

Oberon also saw this as a characteristic that divided people into two groups: "Whenever I teach people to use computers, I recognize them as being in one of two groups - they are either afraid of breaking the computer by doing the wrong thing, or they aren't. The ones who aren't don't really need a teacher, so much as a guide." It seemed to be that this sample was not afraid to make mistakes or worry about breaking the computer.

Even though they felt they could easily recognize if people had volition or not, they admitted that they did not always put effort forward themselves. Oberon confessed, "Oh, I am a total techno slacker. Once I was forced to write 1000 lines as punishment for some public school nonsense, and convinced my teacher to let me write a program that would write the lines for me in BASIC and print to the Commodore dot matrix." If there was a way they could use technology to make a task quicker and simpler, they would do that. Sometimes it seemed easier for them to do low-level work and stay comfortable, but in the long run it was not satisfying for those who have real talent. Escher talked about his laziness in relation to technology, and how he eventually had to quit his low-level job:

101

Oh, I was really lazy. Like I got my job at [greeting card company] while I was an undergrad, working part-time, and I graduated and of course there weren't a lot of IT jobs available in 2001. So I just kinda hung out at [greeting card company] where I had a job and things looked good. But, again it wasn't challenging, it didn't pay great. It wasn't challenging, and I was a total slacker. It was terrible.

Despite describing themselves as lazy, if these participants cared about a project they would not hesitate to work long hours on it. Escher also said, "I know such a variety of people. It seems to be a common thread that they're people that like to stay up all night long." Escher admitted to being one of those "night owl" type people himself in high school: "I remember my dad, my dad was real flexible with me, I could do anything that I wanted. But staying out til 4 AM and then going to high school the next day, he didn't want me doing that. I'm like, 'But dad! I'm learning!'" Therefore they made the choice to be lazy in other areas of life so that they could focus their effort on working with technology.

Self-Management.

This sample talked about initiative, curiosity, and concentration as aspects of self-management they thought were important. The participants talked about curiosity as something that spurred them to initiate activities on their own. Hugo advised kids to make that effort: "Start as early as you can to learn about various aspects of programming and computers. Don't just accept the learning resources that appear to be available immediately - if you have an interest that isn't being satisfied by your current resources, look for resources that will satisfy your curiosity." Escher remembered in his elementary school that he took it upon himself to learn how to program in the library:

I wanted to use the computers at school more, because they had a lot of Apple IIs. And I'd check out books from the library and write, like basically from the appendix of the book, type in the program and try to run it. Of course I made so many typos it didn't work, and I really didn't get very far that way, but I don't know, I kinda got an understanding of how programming worked.

Felix also wrote about what he thought were the most important qualities for technology talent: "First and foremost they need to have the ability to learn. Part of that, but also secondary to it, is the ability to get information they don't already have. Love of technology itself will only go so far, but these two skills will take a fledgling technologist farther than any others."

The ability to concentrate and stay focused on a task is another aspect of self-management. That focus can sometimes lead to tunnel-vision, and a balance can be found. Felix admitted, "I prefer formulating ideas in groups but doing the actual work on those ideas by myself, because I find that often I stick to one idea and don't really pick it apart or consider alternatives as thoroughly as I ought to." Also sometimes they can become focused on too many things, and that can cause more stress on the individual. Hugo reflected, "I don't mind helping others when they have a question about something, not just limited to computer problems (though that is certainly where I excel). My only problem then becomes that I tend to spread myself too thin, taking on too many problems other people have at once."

But they had an ability to multi-task that was not mentioned in the historical sample. Again it was Felix who stated, "I am currently answering emails, serving mp3s via an FTP daemon running on both of my computers together, reading web comics and downloading a *Flight of the Conchords* episode." Whether this lack of focus on one task at a time was positive or negative was not discussed. However, the people in this sample seemed to enjoy being able to do many things at once with technology. Argus wrote:

103

My entire life is computer-based. I use computers and technology for most of my life. I make a living off of it, doing professional software engineering. I also use it to make music and mix CDs. I use it as a typewriter when I'm writing. I use it for research. There is hardly a part of my life that isn't significantly touched by computers and technology.

He structured his work and leisure around technology, and was focused on the activities that interested him. This sample was almost entirely self-taught, and they could concentrate on multiple technology projects at once with ease.

Personality.

Overall with this group, there was an expression of rebellion and statements made in relation to being a non-conformist. Hacking was an activity used against authority and the establishment. Felix remembered one particular incident:

I also remember hooking the 'portable' computer up to a telephone using magic (actually an acoustic coupler, but at the time it was magic) and watching my dad program the machine to war-dial a televangelist's (either Falwell or Swaggart, I am pretty sure) 1-800 number in order to run up as much of a phone bill on their end as he possibly could.

The feeling that all rules could be broken with a computer played a big part in their personalities. These traits were similar to ones observed in students in a study by Braun (1999) titled, *From Misfits to Software Pioneers*. The participants in this sample also have a quirky sense of humor, similar to other samples. Escher related how he experienced his sense of humor in two different environments:

I came up with something to say a long time ago, if somebody said 'What's up?' and I'd say 'Half of the universe.' And nobody at [greeting card company], they'd be like 'What, that doesn't make any sense? What is that? Can't you just say 'not much'?' And anyway, so one time I said that to an engineer at [university], and they laughed, they laughed hard! It's not that funny, but I felt like now I'm in the right spot.

Personality was the area that had the most variability based on individual responses. In relation to temperament sometimes they reported that their frustrations could become overwhelming, as Felix demonstrated, "I am personally given to fits of very short temper when it comes to technology, more so over the last few years and tend to do rash things like throwing laptops across the room when they do not do my bidding in terms of compiling and running software, that sort of thing." Others have a much more calm and easy-going temperament, like Hugo: "I've always had trouble gauging what others think of me, but I would guess they would say I am reserved, quiet, generally pleasant, but not overtly or actively sociable or friendly." When asked to describe his personality, Slate simply wrote, "Mega-nerdy/geeky." Whatever aspects of their personality they chose to emphasize, they all made it work towards their technology interest.

The participants in this sample placed a great emphasis on the importance of intrapersonal qualities like autonomy, curiosity, and persistence to technology talent development. As Argus stated, "I think that being an autodidact is more important than being intelligent because to be autodidactic one has to be intelligent but the inverse is not always true."

Environmental Catalysts

Family.

When talking about their families the participants mostly focused on their role as acquirers of computers and technology for the home. Oberon wrote:

> I remember using an electronic typewriter in my father's office that had a tiny 3-line LCD display where one could compose the text of document files before the typewriter printed them out. Our first home computer was a Hewlett Packard x86 IBM clone with a monochrome monitor.

Access was not always directly from parents, and other family members introduced some of them to the world of computers. Hugo remembered, "When I was in fifth grade, an

105

uncle bought my family a Commodore 64 computer which I immediately began to use to play games and soon after to start learning how to program software." The simple act of introducing a computer to the participants was enough to spark their interest. Some members of their family also shared their technology enthusiasm. Pike stated, "When I was six, we bought a Commodore Vic 20 with a tape drive. It was fun but unreliable. My cousin soon bought a C64 that was much cooler.... I always thought they were really cool, and my family seemed to agree." At this period in time around 1982, it was possible for a child to ask for a computer and for their parents to be able to buy one for just under 600 dollars, (*Computer History Museum*, 2006). Argus remembered, "My first computer was a Commodore 64 that I received for Christmas when I was 10 or 11. My family was very supportive and understanding."

Four out of nine participants mentioned that their parents were a strong influence on their early experiences with technology. Felix wrote that his greatest influence was:

> Definitely my family, more specifically my dad. I am told that in elementary school I wrote an essay on what I wanted to do when I grew up that essentially read 'I want to work in computers like my dad. I will work with computers, wear a suit and make lots of money.' Two out of three isn't bad.

Escher said he was strongly influenced by his father who would often sit at the computer with him when he was young:

> One of the games that my dad got on our PC was called *Jet*. It was a flight simulator of F-16s and F-18s, and that really tied in closely with my interest in airplanes. So I'd sit on his lap, and I was kinda the bombardier. He'd fly the airplane and I'd push the space bar to drop the bomb or something.

For Sako, his father purposefully guided him through the world of computing, as described in this quote:

> My father provided me a computer back in 1978. This was well before home computers were common. It was an old Altair. Through childhood he actively

provided newer equipment, books, electronics supplies, and access to university computing equipment and facilities. I was on the Internet (the ArpaNet as it was called then) in 1984…. Further, I was interned as a child to companies that my father had contacts within well before such things were probably legal.

Slate indicated that he was not directly guided by his parents as much as influenced by their own technology interests: "My dad's an engineer who used to work with Steve Wozniak, and my mom worked at the NASA Ames research center, so I've definitely grown up nerdy. I was brought up a techie." Both his parents were early adopters of technology so it seemed natural to him growing up in that environment. Having supportive parents in general was helpful to some of the participants. Felix wrote, "My parents have always told me that I'm a smart and capable person and how proud of me they are." Unfortunately, not every participant had a positive experience with their families. Moby wrote:

> I'm looking back on that then and saying to myself, 'If I were my parents then I would have done this and this to encourage me.' It was really something they didn't understand at all. My mom still can barely use the computer, but I don't think you have to understand it as I do to encourage.

He stated that this contributed even more to his self-reliance, but he would have liked more support at home of his technology interests.

School.

All participants had had some experience with computers while in school, but this was not their main source of learning. Two participants mentioned taking computer classes in elementary school, a curriculum addition that gained momentum in the 1980s. Pike stated, "In grade school I learned to program on the Apple II, but all we really made were silly moving graphics…. I had computer classes in 3rd or 4th grade, and word processing in 9th or 10th." But it depended on the school's resources. Felix reflected:

107

There were computer classes around that time but nothing more in depth than using Logo (turtle down, forward 100, turtle right, forward 100, etc). I don't really remember much else computer-wise until high school honestly, but I also moved around a lot and was thrown into different systems which were not as technologically advanced because they couldn't afford to be.

The problem with the schools at this time was that they were unable to keep up with the pace of technology. Hugo wrote, "By the time I was in high school my experience was already beyond the computer classes that were offered so I opted to take art classes instead." The students already knew more outside of school, so there was nothing they could gain at this level.

Only one respondent clearly remembered individual teachers who inspired him with computers. Hugo stated his greatest influence was found at school:

When I was in fourth and fifth grade we had an Apple II computer at school that our teachers would use to familiarize us with the computer and also to teach us by using educational software.... Mostly it was teachers. I remember specifically two teachers that were the biggest influences when I was first starting to learn about computers.

Another respondent also had a positive experience in school, and he felt being in the gifted program was helpful. Oberon reflected:

I remember using Commodore 64s and Unisys Icon terminals in the public school system in Toronto. The Icons were in 'regular' schools at the time and Commodores were at the 'gifted' schools, where I was.... I was in gifted classes, which in Toronto had different computers and also, in high school, did take a public school computer programming class that focused on QBASIC.

Most of the respondents reported that they had neutral or negative experiences at school, and did not think they gained much from the classes at the time. Argus wrote:

The only computer courses I've ever taken were the ones forced upon me during Junior High. It was all based around the Apple IIe and all I can really remember doing was using Print Shop to make greeting cards for no reason whatsoever. It was mostly a useless class. We'd sit in the back and secretly play *Karateka*.

What was offered at the high school level was limited to word processing and keyboarding, and some basic programming. Moby wrote, "I took a bit of crappy BASIC and Pascal programming in high school." After his negative experience, Moby offered a suggestion to help gifted students like him:

> For me, it is important to have other kids around. It doesn't necessarily have to be all technology kids at once. But there should be a technology 'corner' or section that these kids could have in a gifted classroom. Where they could have the time and freedom to follow up on their own questions about technology. That's what is different from the normal education system letting kids have freedom and responsibility for their own learning.

Overall, the participants expressed the opinion that school was a waste of time, and that the resources available to them at the time were inadequate to advance their learning in this area.

Peers.

Five out of nine specifically mentioned friends as an influence in developing their technology interest. Computers were more prevalent at this time, compared to the historical sample, so it was easier to find peers with the same interests. When asked how to identify students with technology talent, Oberon wrote, "Mostly they are self-identifying. Put a box in the room, and all the nerds will be drawn to it and start to geek out and talk about their own b0xen.[5]" They enjoyed working with others and sharing ideas, but only if they felt the other person was on the same level as them. Slate wrote, "I don't mind helping some people with computer issues, but I *hate* when people think I'm their personal IT department!"

Two participants remembered how important forming a core group of self-styled computer nerds was to their development. During the early nineties, their goal was to

---

[5] The word "b0xen" is a slang term for a computer terminal.

connect and talk to people on-line. Argus was happy because, "All of my friends were into it, too. We were a collective of progressive BBS-running nerds. It was awesome." Being electronically connected to friends was a great motivator to learn more. The focus became how to work together to get access to internet connections, by any means possible. Oberon talked about how, "Some friends and I in high school founded the 'computer science' club as an excuse to get privileged access to the school's networked computers to play games and pirate software. We later founded the school student BBS for basically the same underhanded reason. Good times."

An important aspect related to peers was the dissemination of knowledge and sharing of resources. Almost every participant was excited when he could share ideas with fellow computer enthusiasts. Escher got exposure to new technology through friends:

> I was really learning about ray tracing, and some of my friend's cousins were programmers at [computer company] when they were in Topeka.... So they were coders working on ray tracers, and I was just itching to learn how to use the ray tracers. So I'd go over to their house and use some alpha version, and learn how to do some computer graphics and 3D rendering. And that was really intriguing stuff.

With the advent of the internet, now one's peers did not just mean friends within driving distance. They could found peers all over the world, and used this virtual community to exchange ideas. Moby was an early chat enthusiast: "I did IRC (Internet Relay Chat) for a long time though. I was even on I-link. It was an online service offered by Tandy, I think. I was 13 at the time, so it was 1989. I lied and told everyone I was 31." The desire to learn more and share ideas is part of the culture, and Sako mentioned he was a member of the "Free and Open Source software community." Learning from others was encouraged among this group of participants.

Again in this sample, a specific dyad was mentioned between Escher and Chuck. Escher described Chuck as the mathematical / technical one, and Escher was the creative / artistic one. This partnership benefited them both because they could work together and play off each other's strengths. Escher talked about one collaborative project:

> When I met Chuck, it was like I really had somebody who I could bounce ideas off, and create new things. When he created something new, I would use it, and tell him how to fix it, when he had his bulletin board set up, I would do some artwork for him, you know, ANSI artwork. So he taught me what ANSI was, and gave me the editor, and I basically put together the scary monsters and funny fonts and things like that, and put that on his BBS.

Whether it was one solid friend or a big group of on-line contacts, peers played a big part in the technological lives of the participants in this sample.

Media.

Three participants specifically mentioned how media influenced their interest in technology, which was a perspective unique to this sample. Sako enjoyed the surge in computer-related media: "Fictional movies, such as *War Games*, helped with my imagination. Computing journals, *Dr. Dobbs* is a great example, helped fill in some of the early gaps." In the eighties and nineties, images of new computers filled movies, television and magazines. Technology was featured everywhere. When Argus was asked what the biggest influence on his technology interest was, he responded:

> The media and video games, really. I was at the perfect age when home video games and computers broke out into the mainstream. There were arcade machines in every grocery store. I was perfectly poised to take on the entire wave whether it was moves like *WarGames* and *Tron* or TV series like *Whiz Kids* and *Automan*. I ate it all up.

When asked if he was influenced by any movies related to computers, Oberon wrote, "For me, it's a toss up between *Flight of the Navigator* and *Tron*." Argus talked about one very distinct memory he had related to computers and media:

> I wanted a computer originally only because I liked the sound of the keyboard. I was on vacation with my parents and we were in a hotel room somewhere watching a TV sitcom wherein someone was typing on a keyboard. I liked the sound so much, I asked for a computer that year.

All of the media images being promoted related to computers and technology caught the eye of children with that curiosity and inclination to explore. Knowing the resources were out there helped them to actively seek out technology opportunities.

Talent

As with the historical sample, the way in which most participants expressed their talents was through their work. However the career requirements had changed from the 1980s when the first programmers were interviewed. There were many more options where a person with technology talent could use their skills. Sako was one of the participants who tried many different types of jobs:

> I currently work as a tech for an engineering reprographics firm that involves supporting business customers along the entire gambit of their software / networking / hardware needs. I have been employed in technology training, software development, system administration, networking, support, documentation, and consulting roles. These employments have been across the spectrum of internet start-ups, educational institutions, small business, and enterprise level organizations.

There were multiple options related to software, hardware, networking, and technical support where they could find employment. The level of responsibility ranged according to individuals, with some choosing low-level jobs and others heading large projects. Felix had an influential position at a networking company: "I am currently leaving a job wherein I helped design an international IP backbone for a fiber optics company, and helped to create new products based on upcoming technologies in the metro Ethernet market." Some of the other participants, in line with their personality and level of motivation, chose jobs where they could complete tasks quickly without having many responsibilities to the company. Argus described one such job he had:

> When I worked at [software company], I wasn't much more than a low-level coding monkey....They would just hand me an object model, a list of tasks that already had time allotted for everything, and told me to 'go.' I always had my work finished with plenty of time left and it was an absolutely perfect work environment for me. They said, 'Do this in four hours,' and I said, 'Okay,' and had it done in two.

He said that he just wanted to clock in, do his work, and clock out, and that work was not where he expended most of his energy. From his own observation, Sako determined that, "Honestly, not very many of the people who try to become professionals in the world of technology seem to be very happy." Very few of those interviewed in this sample felt that their job was what defined them as an individual.

Besides having a technology career, the participants in this study also used computers in a variety of ways at home. They created multimedia projects, chatted with others, posted to message boards, did some troubleshooting, and played games. Argus spent most of his time outside of work creating electronic music using a computer keyboard as his instrument. It was in this aspect that his technology talent was most apparent, and he was able to express his creative abilities. Moby was another talented musician who incorporated technology into his compositions. When asked how they used technology at home, most of the participants talked about using it for entertainment and communication, as well as working on their own personal coding interests. Pike enjoyed using computers: "I mainly use them for communicating and networking with people. I also use them for gaming. When I need a new computer, I buy the parts and build it." He used his skills with hardware when necessary, but mostly he would just play with computers. Slate was another participant who focused on games and communication: "I also game on the PC in my spare time and maintain a web server at home for personal code projects." For him, software coding was a fun pastime and he did it for the fun of creating and sharing his ideas. One final way that they used their talents was in helping other people with technology problems. Felix wrote that, "I am actually extremely fond of helping people with computer problems that I don't yet know the solutions to, because I find we both end up learning a lot in the process." The degree to which each participant enjoyed helping others varied with their temperament and tolerance for other people.

Oberon joined the 8% of Americans who were classified as 'lackluster veterans' and was no longer thrilled by computer technology and internet connectivity (Horrigan, 2007). After years of being one of the most involved hackers in this study, he quit his technology job and decided to become a butcher. When asked about his current use of computers, he responded:

> Nothing, thank god. Other than keeping a 6-year-old computer I cobbled together from bits I found in the trash running Windows XP well enough to watch Google video and post to LiveJournal, and occasionally design flyers for the butcher shop in InDesign.... The point I was trying to make was not that I'd given up on being a techie, but that I'd given up on being a computer techie. Now I run machines (grinders, choppers, etc.), I collect *Star Wars* droids and I hack life. I study theology and science, and I look for truth and enlightened reality. I use a computer (and the net) to do it, but I've managed to push that aptitude out of my core identity into the tool category like a toothbrush or an automobile.

He was unique to this sample because he made the active choice to reject technology and follow a more philosophical path in life. There were still aspects of technology that he used for entertainment and to communicate with friends, but it was no longer something that defined who he was as a person.

I felt it was important to include his perspective as a product of talent development, because he went through the entire process and made the decision to follow a non-technology path, even though he had some of the highest natural abilities. In relation to other areas of talent development, Bamberger (1982) studied musical prodigies and found that some of them suffered what she called a *midlife crisis* during adolescence or young adulthood. These precocious youths reached a point in their development where they faced a choice between continued specialization or changing their focus in life to another area of expertise. Some of the individuals she studied simply "burned out" on music and stopped playing an instrument entirely. Oberon's description of how he abruptly stopped using computers and chose a different career echoed Bamberger's findings. It is up to every individual the extent to which they infuse their life with computers and technology, and that aspect of free will should not be overlooked.

Predictions for the Future

Of the four samples interviewed for this study, the participants in this group seemed to have the most pessimistic point of view in relation to the future of technology. Oberon stated that society was going, "Straight to hell. We are destroying our capacity for genuine human interaction." Argus agreed and predicted that the future was heading in a bad direction: "We're already generating children who are emotionally and socially retarded. We can't even memorize seven digit numbers anymore. We rely on technology too much and it's making us soft and stupid." Both of these quotes reflected their disillusionment with the promises that supporters of technology had made during the 1990s. Moby expressed one of his own fears about the future of society:

> The worst thing I think that could happen is the potential for creative or new ways of thought to be inhibited because of our interaction with computers. Computers started out as a new way of life, a new shiny future where everything was chrome plated. I think we must get away from that vision soon and adopt one that's much more healthy.

One participant even said that he thought a big conflict was eminent in relation to technology. Felix wrote: "Perhaps there will be a striation in society wherein luddites join forces and rebel against the technocratic elite." Overall, they did not have many positive things to say about the future.

However, a few participants took a more neutral position in relation to the future of technology, and acknowledged both the positive and negative possibilities. Hugo recognized the historical cycle of innovation, and looked at computers from that perspective:

> Technology will likely continue to become more and more involved in every aspect of our lives. Just as it has since the industrial revolution (and really, since the beginning of time) new things will come along, excitement for them will last a while until the novelty wears off, then it will become commonplace and seem more just like any other aspect of life we take for granted. In the end, how technology affects our lives is more determined by us as a society and as individuals. People will always find ways to use technology for their own ends,
> be it philanthropic or misanthropic.

Slate seemed resign to the fact that technology was going to continue to move forward, despite the consequences: "It's already essential to the majority of the population. I expect it to spread further as time goes on." One of the few participants in this sample who still showed excitement for technological innovation was Escher. He animatedly talked about the capabilities of new 3-D programs and the possibilities of software in the future:

> Visualization of data, and the marching cubes algorithm, I mean, and they're coming up with new stuff that's really neat, and I think that's just fantastically valuable stuff. We're going to be able to understand more and more phenomenon that are just invisible to us…. Just really intensely complicated problems, but if you look at it in the right way, then the human mind, that creative part, sees patterns, and refines the process, and then sees more of the extent of the patterns.

For him, there was still much to be accomplished in the area of research with technology, and he liked to focus on that aspect of it instead of the entertainment value of computers and multimedia.

Even though the majority of this group took a pessimistic stance on the future of technology, they still offered helpful suggestions for the next generation of techies. Oberon shared his philosophical perspective: "Keep the machine invisible. Reality is in the other people looking at the box, and not the box itself." It was important to him that children remember to interact with other people and not just computers. Felix addressed a different issue: "Do remember to go outside as much as possible. You cannot live inside the machine, nor should you try." There where many other experiences in life that cannot be had through a keyboard, and Felix wanted to impress that on the minds of younger people. Argus also encouraged more activities, both physical and mental:

> Go outside, you pasty, chubby runts. Run around. Ride a bike. The computer and the internet will be there when you come back. Memorize your friends' telephone numbers. Memorize your social security number. Do something, anything, to exercise your entire mind. Don't rely on computers for everything because one day you're going to lose your PDA (Personal Digital Assistant) or the lights are going to go out.

Overall, they wanted the next generation of children to experience all aspects of life, and then they could make an informed choice if they wanted to pursue a career in technology or not. The participants seemed jaded that the youth of today took too many shortcuts in life and did not put in the time and effort that they had to do when they were younger.

Hugo had one last final piece of advice:

> Remember that computers and technology are just one aspect of life. Our humanity is the most important aspect, and it's too easy to get lost in technology and forget the rest of the universe exists. A well balanced, well rounded life will ultimately provide for a better existence.

Their generally negative predictions for the future of society did not prevent them from hoping that the next generation of technologically talented children could someday learn to find a balance for all aspects of their lives.

Summary

There was a greater variety of experiences self-reported by this sample than the historical sample, perhaps because they were products of a different generation. They discussed the importance of intelligence and intellectual abilities, but emphasized that just knowing a lot about technology was not necessarily an indicator of talent. To them, it was more important to have creative thinking and an imaginative perspective when approaching problems related to computers. They enjoyed expressing their creativity through technology, and several of them created digital music and art as a hobby. Most of them felt competent with their own socioaffective abilities, but acknowledged that the stereotype of the awkward nerd was still prevalent in the people whom with they worked.

As a whole, the self-reported motivation of this sample was lower than the first historical sample related to school and work. They used technology to make things as easy for themselves as possible, and called themselves "lazy" when it came to doing work that did not interest them. In contrast, when they worked on their own computer projects, they would expend a lot of effort and work long hours and late nights. This quality reflected their non-conformist personalities, and that they would not be forced to use their technological skills unless they wanted to. In relation to external influences, peers still played a large part, as well as the media and some family. More access to different types of computer technology were available as they were growing up, and so they were able to explore on their own more. The resources at their schools improved with time, but the skills of the individuals in this sample continued to outpace what was

118

taught in the classroom. When asked about their predictions for the future of technology, they gave a generally pessimistic point of view. Sako offered his own philosophical take on what technology means to society:

> Technology is the practical expression of scientific humanism. Being a 'techie' is more about developing a way of thinking, understanding yourself, and perceiving your world. It is not about video games, and YouTube, and the iPod. Those are just popular expressions of the need to create and share.

Finally, they offered recommendations to the next generation of computer users to go outside and not live their entire lives through technology, which seemed contradictory to what many of them said about their own lives.

Sample 3: Longitudinal

For this sample, there were two interview times where data collection took place.

At Time 1, the participants (N=9, Male=8, Female=1) were involved in a pilot study

where they created a website for the project sponsor, Dr. Turing. They were asked

interview questions upon the completion of the project, and data were collected through

transcripts and Dr. Turing's handwritten interview notes (Appendix C). Data for Time 2

was collected four years later, when many of the students were completing college.

Table 3.

Information Table for Sample 3

Name	Major	Gifted	Age	Location
"Mark"	Mechanical Engineering, plans to get a Master's degree in Computer Science	Yes	22	Chicago, IL
"Janus"	Started in Engineering, switched to liberal arts school for Pre-Law	Yes	22	Lawrence, KS
"Lisa"	Philosophy, plans to get a Master's degree in Library Science / Information Systems	Yes	21	New York, NY
"Able"	Electrical Engineering and Computer Science, programmer for university computer project	No	22	Lawrence, KS
"Haskell"	Political Science, plans to study Cryptology and go into national intelligence	Yes	22	Selinsgrove, PA
"Pascal"	Computer Engineering, database designer for the Molecular & Cellular Biology department	Yes	22	Springfield, IL
"Franz"	Mathematics and Computer Science, plans to work for Google computer company	Yes	22	Northfield, MN
"Rexx"[a]	Listed as Psychology major on Facebook.com	No	22	Lawrence, KS
"Tom"[a]	Unknown	No	22	Unknown

[a]Participated in pilot study but did not participate in follow-up questions.

The participants (N=7, Male=6, Female=1) were contacted electronically to find out how they had pursued their computer technology interests since high school. The students who agreed to participate responded to questions through e-mail (Appendix E). Their age at the time of the second interview, location, college major, and whether or not they received any gifted services are presented in Table 3.

Natural Abilities

　Intellectual.

The intellectual abilities emphasized by this sample included intelligence, logical thinking, and a wide base of knowledge about computer technology. In high school, six out of nine were officially identified as gifted under the state and district guidelines. Being part of the gifted program supported the assumption of high intelligence in these students. However, participation in the computer programming club was voluntary and based solely on interest. No matter what their level of measured intelligence was, they all had a desire to learn more about computers and programming.

Some of the participants discussed the importance of intelligence as the starting point of CTT development. When asked about computer technology talent in a follow-up interview, Janus wrote:

> If a kid is intelligent and interested in computers, he should be given a productive framework and resources, like access to more sophisticated systems and powerful software the same way a gifted kid who is interested in say the French language should be given opportunities to be immersed in French culture.

He thought that with the right resources, an individual could expand their level of knowledge and skills in technology. This view reflected what Janus originally said in his first interview: "I would like to work on the cutting edge – with the equipment, high powered computing (e.g. exotic machines), something mentally challenging rather than

121

labor intensive." It was important to him that he could use his brain to solve complex problems in his career. Dr. Turing, an expert in gifted education, observed the way this group demonstrated abstract reasoning while working on the website project: "Manipulation abstract symbol systems re computer language - some of them even learned a new language to do the programming for the web site." She concluded that overall these students had high intellectual abilities and were able to comprehend complex systems and languages quickly.

In relation to intellectual ability, the process of mathematics and logical analysis was discussed. At the time of the follow-up interview, four participants were in the process of completing math or science-related college degrees. Their amount of interest in the computer science aspect of the project varied by individual. In his original interview, Tom stated, "I've got a pretty good mind for logic, so I can see how to use programming to solve a problem, I can sift through information to solve." He also rated himself high on numbers and spatial skills and said that, "Generally in a math class I'll pick up things in a day." Unfortunately, the researcher was unable to locate Tom for a follow-up interview to find out if he pursued this interest in math in college. When asked how he would distinguish technology talent students, Mark wrote: "They'd be identified in all the ways you would expect, gifted at math and science, playing with electronics, etc." Not only excellent mathematical thinking but also the desire to explore computers and technology should be present, according to him.

Another factor mentioned by this sample was the desire to have a deep level of knowledge. Able stated, "There are certain people who just click with technology. It may be beneficial for them to receive access to new information and ideas. Things are always changing in the technology field, and if you can't keep up, you will quickly fall behind."

The amount of information a person knows was considered less important than keeping up-to-date with current technology advancements. Franz found this aspect tricky to assess and said it would be difficult to test for this knowledge in a systematic way:

> I don't think it makes sense to create some sort of static test that quizzes students on their computer knowledge (e.g. 'What is a processor and what does its function?', 'What are some popular languages used to create and run websites?'). However, if you could create some sort of interview process, through which you could determine that a students demonstrates a desire and ability to think about technology on a higher level, that might be more appropriate (though, admittedly, also difficult to create / grade).

Having only a test of knowledge would provide limited results, according to Franz. Overall, intelligence, mathematical thinking and a desire for technological knowledge were the intellectual abilities mentioned by this sample.

Creative.

In relation to the area of creativity, this sample mentioned artistic ability and imagination most frequently in interviews. Four out of the original nine participants specifically mentioned working on some sort of creative outlet using technology. Some students were clear about their limitations, like Mark: "I don't have that artistic sense. I'm not necessarily artistic." Even though he didn't think of himself as an artist, he still produced creative media projects using, "Special effects, motion animation; film media classes – makes movies on computer." Other participants also talked about their creative perspectives. Rexx said his strengths were in, "Graphics – layout, I guess I'm more of an artsy person." He enjoyed working on movies with Mark in and out of school, and their abilities complimented each other. Janus also created movies and graphics, and talked about how he learned to use Photoshop on his own. Lisa said she enjoyed graphical design, and she used her talents to create business cards for clients. As an educator, Dr. Turing noticed that, "The kids who worked on the graphics also demonstrated visual thinking - in three dimensions. Planning, forecasting, sequencing." Not only did they have an artistic sensibility but they had the thought processes to carry out and complete their projects.

There was another avenue where these students discussed using their imagination, and that was in the area of video games. Tom said he enjoyed, "Role playing video games, tell stories, *Final Fantasy* – I see them more as an artistic medium, beautiful artwork and music and tell wonderful stories." Eight out of nine of the original sample said they preferred strategy and role playing games where there were multiple paths to take. Haskell stated he didn't like chess because it, "Seems limited – prefer games that have almost infinite options – NOT certain way to win/lose." They liked having open possibilities in games where they could imagine different strategies for success. Lisa mentioned she played the *SIMS* online because she liked building the house and accessories, and she also enjoyed the on-line community that surrounded the game. In that game, there is no end the users can continue to build objects and interact as long as they like. They did not like restrictions placed on their imagination, in video games or other projects. This group valued creativity, whether they considered themselves artistic or not.

Socioaffective.

In the pilot study, the original nine participants were required to work together to produce a website for the project leader. Their abilities to communicate with each other and collaborate successfully was observed by Dr. Turing, "There were considerable

124

differences in how well the kids worked together. The best programmers were not the best people managers. They found it nearly impossible to let go of any interesting aspect of the project - they wanted to do everything themselves." The group worked because it had a mix of students with leadership abilities and technical skills. After the pilot study was complete, she asked them the question, "In what ways do you imagine that you could use what you've learned from working on this project?" Six out of nine students said they learned communication skills and group dynamics from this project. For example, Tom said he learned about managing a group project, and how to balance between "talking and doing." He thought they worked together successfully when they broke into different groups that focused on each person's talents. As a team, they delegated work to either the programming group or the website design group and worked toward their strengths. They saw how communicating their ideas to other people was essential in a work environment. It was also helpful to them to see how to handle conflict between themselves. Dr. Turing said, "The kids who defused challenging interpersonal situations were not the ones who parsed the project and sequenced the tasks." For most of them this was the first time they had completed a technology group project together, and they were thankful for the social knowledge they gained.

As for the social stereotype of the awkward nerd, it was hard for the researcher to determine this aspect solely from their responses. Janus admitted, "I didn't fit in during junior high and high school," but he said he found people to interact with in the programming club. Lisa jokingly said that she joined the computer club to meet guys. The participants did not discuss their social relationships outside of their high school's Computer Programming Club, and then seemed more comfortable with peers with the same interests. Some did fit the stereotype, as Dr. Turing related: "These are kids whose idea of a social event was a LAN (Local Area Network) party in which they sat in the same room with their computers networked together and never even got up to visit in real time." However she said the majority had a normal level of social skills, and some were even considered popular: "Several were scholar athletes and at least two of the guys were either homecoming or prom kings." This supported the idea that computer interest was more accepted by peers as a mainstream activity. In the follow-up interview, Haskell was asked his opinion about creating a group for computer technology students, and he wrote:

> I would support such a program. While it is much easier for kids these days to 'admit' enjoying computers and technology, fostering an encouraging environment for kids to develop technology skills while also giving them social connections to others similarly gifted without fear of harassment could only benefit these students.

Although this group was able to communicate with each other, they recognized that a safe environment was provided for them in the club and thought that similar clubs at other schools could prove to be beneficial.

Intrapersonal Catalysts

Motivation.

There was a mixture of internal and external motivation that helped this sample towards their goals. There was the immediate goal of completing the website project, and all of the participants said they were satisfied with the end product. Tom gave the project a rating of 8 out of 10 because, "A lot of what we set out to do was accomplished." Pascal thought they did very well for never having worked like this before: "I believe this was a very successful project. Group's first experience, actually completed it as envisioned it, actually split up the work, pieced together nicely." There was the external pressure of a timeline and specific tasks that needed to be completed, and some of the

126

members worked harder than others towards those goals. They were able to balance the workload between themselves so that everything was eventually completed at the end.

Some of them said they would use their skills in the future, but others had not figured out what they wanted to do with their lives yet. Able knew he wanted to pursue technology: "I now have more experience with making an actual website that people will use. This experience can help me with my future education and career." On the other hand, Haskell said that he would continue to keep up some basic computer skills, but that he did not see pursuing technology in the future. Janus had a realistic idea on what steps needed to be taken in high school to help students achieve their goals:

> Don't get too absorbed in computers and keep a real perspective on what they are and how they fit into society. Also, if you are interested in computers ask for help from counselors in finding programs and experiences that you can put onto a resume for college applications and that are worthwhile, instead of just 'doing' computers.

Whether it was completing a project or deciding on their future career, the majority of these students were motivated.

These students agreed to participate in this project because they enjoyed using their skills and knowledge and gained satisfaction working with technology. Dr. Turing observed, "Several students expressed pleasure about learning new skills….
It was clear that they were doing what they loved - the tasks alone were motivating.
Completion was not the goal!" Excitement and passion fueled many of the participants, and Lisa wrote, "I loved the feeling that I knew more than grown ups." The enjoyment of working with technology and creating projects was part of everyday life for most of them. During the pilot study, Able stated, "I had already made several webpages for fun. LearnGen gave me the opportunity to make a real webpage."

Along with all those internally motivated feelings there were some external motivators as well. Lisa used her love for computers to start her own business in eighth grade, "Repairing crashed computers for a core group of clients, raise up applications, tutor how to use software." Making money and pleasing her clients were an added bonus to using her technology skills. Franz was another participant who wrote about some outside motivation: "In high school, my main motivation came from my friends and family, whose reliance on my computer knowledge to help them with their technology problems and challenges grew as my knowledge grew." He enjoyed solving problems, and being a troubleshooter for his friends and family (and later *Geeks on Wheels*) allowed him to get validation for his skills. Overall, the participants said they were motivated to learn more about technology out of their own interest and not an external award.

Volition.

This group showed persistence while working on the project as well as other areas of their lives. They continued to pursue their own learning, and Pascal described how he went beyond what was required of him at school: "After learning the basics, I experimented with the computer and developed my own interest in technology." Dr. Turing observed during the project that this quality of persistence turned to stubbornness when, "Two kids tied up the project for three weeks trying to animate the Jayhawk so that it would walk across the KU banner at the top of the home page." Another way that they persevered was working through their mistakes and not being afraid to open up their computers and explore. Lisa encouraged others to be fearless: "Take things apart. Don't be afraid to break the warranty. Go ahead and over clock the system. Hack the planet!" The individuals in this sample emphasized the importance of exploring beyond the

128

requirements to learn new technology skills.

What often came with persistence was the willingness to work hard on problems and not give up. Mark described his high school experience as filled with "nonstop use" of his computer, and estimated working 20 to 30 hours a week on his own projects. The other participants were also lengthy users, but Mark was near the extreme. In addition to expending a lot of effort on computing at home, they also worked on the website project. Dr. Turing noted their pattern of working: "They worked hard in spurts. As the end of the project drew closer, they stepped up the pace - it was hard for me to produce enough raw material quickly enough for them." For those students with the talent, the hard work was worth it. However, some of them had hopes it would be easier. After the project, Janus said that he would like to focus on programming but it was, "Too hard to learn by one's self. With programming there is a lot to do." He concluded by saying that he preferred relaxing and playing video games with his friends. This statement foreshadowed his college education where he started in engineering but switched to liberal arts after awhile.

Self-Management.

Most of the participants considered themselves to be self-taught when it came to computers and technology. They took the initiative to learn more, and looked for answers to their own questions when the people around them did not have the solution. One particularly stellar example was Lisa. She was homeschooled for most of her younger education, and had a voracious appetite for knowledge. She wrote that, "In 5th grade my family got the internet (cable!) and I read a whole book called *The Internet for Kids* from cover to cover. I chatted with people around the world, and I loved it." Pascal was another student who was influenced early by his older brother but then took it upon

129

himself to learn Visual C++ just out of middle school. He recommended that kids should, "Start early and actively seek opportunities to learn more about technology." Although they talked about teaching themselves on their own, when it came to the group project, Dr. Turing stated that, "They were extraordinarily reticent about demonstrating initiative, however. Initially I wanted them to take the lead on web site design, since I didn't know anything about web-based instruction. That set us back for a few months - they wanted me to make those decisions." They were not yet willing to take on the responsibility of a client's work without guidance. This could be attributed to their young age and pressure to do well on an authentic product.

Another aspect of self-management was the level of concentration and focus they had while working. What was interesting about this sample is that they encouraged others not to focus too much on one thing in computing. They encouraged multitasking and getting a variety of experiences with technology. Haskell wrote: "There's always someone doing something new. Instead of spending all one's time programming, try to constantly learn what others are doing. Don't reinvent the wheel, accept what people have done and take it somewhere new." This advice came after he had gone through college and seen what options were available to him. Lisa also encouraged diversity: "There are a lot of fields you can go into if you're good with computers. Don't limit yourself to one genre, think broadly, computers are everywhere, and don't limit yourself to Windows! Try Linux." However, if an individual was passionate about a certain topic, then that person should not be discouraged from pursuing it. Franz offered this statement that emphasized maintaining a balance:

> Many people will tell you to find one thing you're really good at (even within the field of IT or computer science), and develop that skill exclusively. Others will warn

that you must acquire at least a basic level of understanding in a large number of fields (or risk having your very specific knowledge become obsolete very quickly). I would encourage you to do both - make sure to study as many different areas of technology / computers as you can, but don't be afraid to pursue something that really interests you, even if it means that you can't learn about everything else at the same time.

It was important to this group to keep learning about what technology was out there, and to choose what areas they wanted to focus on after experimenting in different areas.

Personality.

With this sample, I was not involved with the data collection at Time 1, and so I was unable to observe the personalities of these students up close. Dr. Turing worked with this group during their website project, and she determined that, "They generally were ordinary kids. I can think of two who were self-described oddballs." There was a range of personalities from outgoing to shy, and it was hard to point to one trait as being characteristic of this group. For example, there were people like Franz who said, "I'm not a very competitive person." And then there was someone like Mark who talked about spending hours playing first person shooter games with lots of action and competition. As with most gifted children, they did express a quirky sense of humor. When asked what technology he would like in an ideal world, Pascal joked, "I currently have rocks and sticks, but in an ideal world, I would like robots, lasers, spaceships, and A.I. (Artificial Intelligence)." There was some non-conformist attitude displayed with comments like Mark's: "Question everything." But on the whole, the personalities of this group were fairly mainstream and normal.

In relation to temperament, there was again a range from even-tempered to easily frustrated, depending on the individual students. The most frequent frustrations mentioned by the participants were conflicts with other people. Able said, "I prefer to

131

work independently, that way I don't have to be held behind or worry about other people's mistakes." Multiple people working on a project could turn into an impediment if they were unable to keep up with the quick pace of the top programmers. Tom expressed a similar sentiment when he said that, "Working with people when skills differ greatly from own, produces high conflict." A different cause of frustration was mentioned by Janus who felt the technology world needed some improvement: "I don't like helping other people associated with business instead of learning the research and science behind it. In many ways computers have frustrated me – too business-oriented and too expensive." Luckily for the pilot study project, there was also a balance of other students like Mark who had good leadership skills and who were able to defuse most conflicts within the group.

Environmental Catalysts

Family.

Three out of nine participants mentioned a direct influence from their parents on the development of their technology interest. Mark stated that his father liked to tinker with all kinds of technology and, "We got our first computer when I was in first grade, my dad encouraged me, he was always a gadget kind of guy and passed this on to me." Haskell remembered how his parents encouraged him with computers, and they taught him how to use the word processor, play games, search the internet and create projects with PowerPoint. In Pascal's case, instead of his parents, he cited his brother as his main influence: "I think my family had the greatest influence on me to pursue my interest in computers and technology. My brother was a computer science major and showed me the potential of computers in the future." Apparently his brother had computer technology talent and Pascal chose to follow in his footsteps. Janus's family influenced his technology interest, but in a different way: "My family was dysfunctional so I absorbed myself in the internet and UNIX system administration lore." Because of the problems at home, it was easier for him to escape into the world of computing. Sometimes a degree of stress like this can actually have a positive effect on talent development. In the article, *Beyond Bloom*, Subotnik, Olszewski-Kubilius, and Arnold (2003) wrote, "Research studies of eminent adults yield retrospective accounts of family environments characterized by stress, trauma, conflict, and dysfunction" (p. 229). Therefore although an unstable home environment is not desirable, some individuals can actually funnel that energy into improving their technology skills and working on their own projects.

Another more indirect way families influenced this sample was through providing resources. In the simplest cases, parents acquired the computers and educational software for their children and let them explore on their own. Lisa wrote about her early experiences with her computer at home:

> My parents got a computer when I was around 8 or so. I played a lot of *Tetris* and *Mario Teaches Typing*. I enjoyed these games immensely. I really enjoyed tweaking the look of the computer, changing the background and welcome message. Later I got into learning tricks such as short cuts which would impress others.

In Able's case, his parents directly encouraged him to learn more about computers, "I have to credit some of my success to my parents. They convinced me to join the computer club when I was in junior high school. I had very little experience with computers at the time." Once he followed his mother's advice, he made friends with other techie kids and expanded his knowledge of technology. Franz had a unique situation because his mother had access to resources at the university which helped him develop his technology interest. At a young age, he was able to receive one-on-one tutoring with a mentor:

> First, I became interested in computers as an elementary school student, and through a sort of ad-hoc mentoring program provided by one of my mom's grad students, I developed my knowledge and interest in computers and the internet gradually through junior high school.

He was able to meet up with a mentor who influenced his interest in computers. Overall, the students in this sample felt supported by their parents in their pursuit of computer technology.

School.

The participants in this sample did not mention their high school as a major influence in the development of their computer technology talent. However, they did say that they enjoyed their experiences in the extracurricular computer programming club. Able wrote, "The only help I have ever received [at school] with computers is from the officers in the Programming Club." To them, it was simply a place to meet at the school that allowed them to share ideas with their friends. When it came to the formal classes provided by the school, the majority of them talked about being disappointed. Pascal explained at length:

> I started in about sixth grade when I designed the elementary school's web site. I then attended junior high and learned on my own about Visual Basic and advanced web development techniques.... However, the school system did not provide a great deal of resources for me to use at school. I took a computer programming class, but the curriculum for both Computer Programming I and II could have been finished in about a quarter.

Many felt the pace was too slow in the classes, and they were limited in the activities they could do on the computers. Mark wanted more personal control and wanted, "To spend a ton of time on the computer exploring.... Typing teachers almost discourage, they're

134

afraid you're going to break it." When it came to using the school resources, the students came up against many restrictions.

Not all of the participants had negative experiences with their school. Tom stated he had a really good webpage teacher in seventh grade who helped his technology development. Dr. Turing was impressed by how articulate the students in this group were about their insights into the public education system:

> Their observations about what schools did/not offer was insightful. For example, they didn't simply trash the school system for its shortcomings regarding hardware, software, or instructional opportunities. Although they were clearly more knowledgeable than most of their teachers, they shared their expertise on request rather than as a power trip.

Able was happy with the level of resources at his high school: "At school I am able to use all of the latest technology that I like. They have Visual Studio .NET on a few computers and Flash on all the computers. I could not ask for more. The technology we have now seems very close to ideal." Several of the high school students were able to attend college classes through the gifted program. Three out of nine learned advanced skills at the university, and Mark took all the web design courses offered during the summer between ninth and tenth grade. Two other students also worked with technology groups at the college, helping students and staff with technical problems.

In the follow-up interviews, the participants said the influence of college played a larger part in their lives. Haskell wrote, "Schools were usually behind the curve, until college when use of email and electronic readings made internet / technology usage mandatory." Some chose to further their education in an area of technology by getting a degree in that subject. Another participant who felt the same way was Franz, "School finally became an influence in college, where I majored in computer science and worked in the Student Computing and Information Center (SCIC), where I assisted students with many technology-related questions." For the students pursuing engineering and computer science degrees, they felt that the courses they took were finally up to their level. As mentioned by Franz, another resource that their colleges offered were technology services, like the student computing center. Janus stated that he worked at one during college:

> My experience with UNIX system administration and computers led me to multiple jobs on campus with programs needing system administrators for desktop support, website setup, database management, etc.

Able was another student who found a technology related job at his university:

> Outside of classes I have worked as a student programmer at the Center for Research and Learning at the University of Kansas. I work with a small team of designers and developers to create small Flash games and applications. I plan on pursuing a career in technology, but I am keeping my options open.

It was also in college where the students explored other options and decided to use their skills in other ways.

Peers.

Not only did this group of students work on projects together in the computer programming club, but they were also friends. For some, it was the only place where they felt accepted during their school years. Lisa talked about what it felt like being the only girl in the computer club:

> In junior high I joined the computer club. Girls aren't expected to like computers, but I loved them. I never really programmed, though I did/do script and in this web based world that's close enough. I was friends with a few people in computer club before I joined it and I felt like I fit right in, as I do categorize myself as geek and frankly, kids in computer club are generally geeks.

Even though the club was school-based, the learning that took place in it was very informal. They influenced each other and would share computer ideas outside of school.

This peer communication was fueled by their desire to learn from each other, as Able wrote:

> The other members of the computer programming club, who eventually became my friends, were eager to learn from and teach their skills to others. Throughout high school, I worked on several projects with these people, and we were always sharing new ideas with each other.

Being part of this peer organization was a positive experience for all of the participants. In other samples, there were clear dyads that formed between a pair of individuals who used their skills to compliment each other. Although this concept was unable to be explored deeply in this group, it appears that at least one partnership was visible between Mark and Rexx. They worked on projects together, and Mark was the technical person and Rexx brought the creative talent with graphics. In the original interview, Rexx said he enjoyed making: "Music videos, graphics, and working with Mark on movies." Other partnerships could have been present among these students, but this was the only dyad specifically mentioned during the interviews.

There were also more casual ways that peers influenced these participants. Mark mentioned that he would have LAN parties with his friends, and they would also "hang out and build computers." Haskell also mentioned gaming as something that brought him and his friends closer together: "As long as I've had a computer with games, I've played them sitting alongside a friend or two. Cooperative play (especially with the advent of ubiquitous internet access) made it all the more appealing." Another place where peers influenced this sample was through on-line communities. Janus talked about the impact the internet had on his technology development:

> I think the biggest influences on my interest in computers and technology would be the internet just because the whole community of computer geeks and Open Source software is online…. A really interesting thing about programmers is their

collaborative approach to sharing software knowledge. I felt at least a part of some social group, even though it became a detriment to my real school coursework and it was ultimately useless on my college applications.

Some students mentioned chatting with online friends and the joy of feeling connected to other people through the internet. Peer influence was important to them, both virtually and face to face.

Talent

Since the pilot study used participants from the computer programming club, that was an area of concentration for all of these students. They also expressed interest in webpage design, troubleshooting, creating digital images, producing multimedia films, posting to message boards, tutoring, and playing video games. There were also some unique ways that they expressed their technology talent, like Lisa who stated, "I'm also the co-host of a community radio station show called 'Technocolor Radio Show' which is a technology talk show." She took her expertise and knowledge to the airwaves and answered questions for people who would call in to her radio show. Although only half of this group chose to pursue technology degrees in college, they still all remained involved with computing. As Haskell wrote: "I remain interested in technology as a hobby (I built my current computer by component rather than from retail) and recreational use of games, graphic design, and geo-spatial applications." This group emphasized that they enjoyed technology in a variety of settings and applications. It was important to them to have choices and not be limited in their exploration of computing, either in their education or their career.

For those participants who planned to focus their adult lives on technology, they liked having choices in front of them. Able wrote, "I plan on pursuing a career in

138

technology, but I am keeping my options open. I like knowing the fact that I could be a network administrator, a Flash developer, or a number of other IT-related jobs." It was again emphasized not to have too narrow a focus, as Franz advised: "I think it is still important that students who are identified as gifted should be challenged in all fields (i.e. not just technology)." This perspective of wide open possibilities could also be reflective of these students being on the threshold of graduating from college. At this stage in life, many young people are still trying to figure out what jobs they would like to pursue, and their career goals might not be solidified yet. Pascal encouraged younger students to get help with their technology interests before entering college to make the transition easier:

> I think during this day and age, many students have a lot of access to technology and already have a lot of potential for pursuing a technology degree in college. I think gifted education should try to provide advanced classes for these potential students such as programming and hardware design or at least provide the necessary material for these students to start studying before heading to college.

He could see the practical ways that computer technology talent could benefit the next generation of students in their education and in their future career.

Predictions for the Future

The majority of the participants in this sample were positive about the future of technology. They recognized how infused computers and technology were in American society, and they predicted that it would continue to spread. Mark stated, "I see technology becoming a greater factor in our daily lives. It is so integrated that we barely notice its presence anymore." Some of them emphasized how small changes would be made to simple day-to-day tasks. Able wrote:

> Technology will make everyday life easer. For example, GPS systems that give you accurate directions to your destination, or a refrigerator that alerts you when you are running out of milk or when the meat has expired.

Janus had a different opinion about the integration of technology. He recognized the advances that have been made in technology learning, but he also cautioned against too much frivolous use of computers:

Computers are so ubiquitous that it is becoming an ingrained part of our existence. Kids are learning how to find resources online at earlier ages. But computers are simply tools for sharing and creating information, and there is a lot online that is a waste of time, like MySpace or chat programs.

Lisa was one of the participants who was looking forward to specific developments in the world of technology: "I'm really excited about 3-D printers and open source technology right now. Collaboratively community created open source physical objects are hot."

Some of the participants looked at technology on a broader scale and how it would affect the entire world. Pascal predicted that computers would become so advanced that they would eventually replace humans in some areas:

Technology is constantly evolving and developing. Many things that are being done by humans will be done by machines in the future. I think in the future, technology will allow humans to concentrate on bettering humanity as a whole than doing basic tasks like farming, manufacturing, etc.

Haskell also looked forward to technological improvements in the workforce:

I believe technology has the ability to vastly improve the efficiency of our service-based economy. Those best able to use the tools of the internet and programming will be best placed for this still-expanding industry.

This positive perspective that most of these participants held reflected their belief that computers and technology were helpful to society. However, one of the students was not as excited about the possibilities of technology as the others. Janus, who switched from an engineering degree to pre-law in college, felt that technology workers would not be as important as other professions:

I don't think computers are some revolutionary development in civilization. They're very useful, and large amounts of data can be distributed, stored, and communicated quickly. IT and technology workers will be an industry suited to maintaining the physical system, but they will never be as valued by society as professionals such as engineers, programmers, doctors, or lawyers.

As important as technology skills will be in the future, he felt that other areas of human pursuit should not be ignored.

Summary

All of the participants in this sample valued intelligence and creative thinking, and perceived themselves as using their abilities to complete work on a group website project. Problem solving, logical thinking and having a diverse technology background were qualities that helped them as they worked together. Their level of social skills varied between individuals, and some were able to communicate their ideas better than others. What motivated them more than anything was an internal desire to explore and learn as much as possible about computers and technology. Many had an insatiable curiosity which was apparent in all aspects of their lives. In their follow-up interviews, they also mentioned how important the external motivators of getting into a good college and finding an appropriate job had become to them. All of the participants were fearless when it came to working with technology, and they put a lot of effort into working on their own projects away from school. They were self-taught, and would sometimes take the initiative, but it depended on the task. There were only a couple of students who were self-described oddballs, and the rest had a pretty normal range of personality traits. This group had a quirky sense of humor that was apparent in many of their responses, and a normal range of temperaments was found among the participants.

Family played a larger influence in this group, and parents bought personal computers and educational software for these students to use at home. Some parents and older family members even had enough technology expertise to directly help these individuals learn about the computer at a young age. Peers were also a strong influence, and they enjoyed sharing ideas and working on computers together with their friends. In relation to school, they had access to some resources but mostly the formal computing classes were inadequate for the level of these students. The school was only helpful in providing a place for them to learn informally in the computer programming club. The talents that these students displayed were even more diverse than the previous two samples, and they would seek out information on any topic that interested them in relation to computers.

Their love for technology was reflected in their overall positive predictions for the future of society. They hoped that computers would improve the everyday lives of people and make it easier for other areas of humanity to be improved. In the end, some students pursued their computer technology interests and others chose different paths, but they all felt their skills had benefited them in their personal development.

Sample 4: Contemporary

The participants in this sample (N=8, Male=8) were all high school students who expressed high levels of interest in computer technology. Pertinent teachers were initially asked to select students who they thought fit the thumbnail descriptions of programmers and interfacers (Appendix B). These students were then given information about the study and asked if they wanted to volunteer to participate. The data presented below were taken from group interviews (Appendix H) as well as chat transcripts and e-mail responses to individual questions (Appendix A). Several parents of the participants (N=3, Male=2, Female=1) agreed to respond to e-mail questions about their technology experience and home environment (Appendix F). The participating teachers (N=5, Female=5) in the areas of media production and gifted education were audio taped and asked questions geared towards them as educators (Appendix G). Information related to the high school participants' age at time of interview, area of technology interest, and whether they received any gifted services are presented in Table 4.

Natural Abilities

Intellectual.

Six out of eight of the participants were either currently in the gifted program or had previously had gifted services when they were younger. This state classification supported the assumption that they had high intellectual abilities. The students in this sample did not openly talk about their level of intelligence but they provided hints about it during the focus group conversations. In the media group interview, Tex said, "I know Jade here is just a whiz. I don't understand half the stuff he does and I don't think I will

143

Table 4.

Information Table for Sample 4

Name	Area of Interest	Gifted	Age
"Cecil"	Programming, hacking, tutoring others about computers, and music	Yes	16
"Dylan"	Programming, hacking, building computers, creating websites, and gaming	Yes	16
"Amos"	Film production, gaming, and a member of Phoenix Films	Yes	16
"Oz"	Multimedia and film production, music, and a member of Phoenix Films	No	17
"Yorick"	Film production, has created numerous *Star Wars* fan films on home computer	Yes	17
"Tex"	Film production, gaming, and a member of Phoenix Films	No	18
"Jade"	Film production and a member of Phoenix Films	Yes	18
"Euclid"	Programming, building computers, and a member of the Robotics Team	Yes	18

ever know." In response to this statement, Jade said,

> I think it's just like you have to be open to learn, and you have to be open for new things. And it's not, I don't know, it's not necessarily that one kid is more technologically minded than the other, I think it's just like a learning ... I think that some people have the brain for it and some people don't, I guess.

He emphasized quick learning and openness to acquiring tech knowledge. When the gifted teacher was asked for distinguishing characteristics, Mrs. Basic said, "I guess their fluency with the knowledge of the materials that exceed the general, same age peer group." So although they did not boast about exceptional intelligence, Mrs. Basic recognized that they were above their peers.

An aspect that was brought up by four out of eight students was the link between mathematical thinking and technology. The logic behind this type of thinking helped, as Cecil explained, "I suppose if I'm trying to figure something out, I try to get a good understanding of what I'm using first. It's important to understand the rules before playing with them." Like an equation, he wants to understand the rules first before breaking them. Cecil openly suggested a link between math and programming in a follow-up quote: "Most of the people who are interested in computers understand programming really well. Actually all the kids that I know understand programming well, they understand math too and understand a lot of the class." Dylan was another student who enjoyed math and technology: "Math and science are fun... I'm good at them anyway. Can't code without math. I take all advanced classes... I'm capable, so I figure that I ought to. A year ahead in math too." During a group interview, Dylan showed the other students a game he had programmed on his calculator.

In relation to inductive and deductive reasoning, this sample talked about the steps they used in problem solving. Cecil stated, "I'll take a look at whatever reference I have or can find if I think I'll need it, then I'll play around with what I know. If something hits me as important as I go along, I'll drop everything else and mess with it." Dylan was similar in his strategy he used to approach a problem: "Just look it up... I guess, and go from there. Try to assemble the skills I need, maybe find a tutorial. But I generally just run on intuition until I hit a roadblock. Online, if I need help with something hard I ask for help." Many of them mentioned finding solutions on-line. Amos used the software to fill in the gaps to complete his work, "I don't think we know like everything, cause there's a ton of stuff that the software can do, but I think if someone came up to us with a certain task, we could eventually get it, and we could eventually problem solve." Sometimes they would be selective about the types of technology they chose to learn about. Yorick, a film media student, said, "I don't know the first thing about how to make my computer work better, but I do know how to make programs on the computer work for me."

Overall this sample expressed intelligence, logical analysis, and deductive thinking as intellectual abilities. When asked how to identify these students, gifted teacher Mrs. Perl said, "Strong problem solving skills, analyzing… I would think kids who are really good at problem solving with the technology are probably good test takers in general." This could be true for the students who already qualify for the gifted program, but some might fall outside of that generalization.

Creative.

Under the heading of creativity, this sample mentioned imagination, inventiveness, and an artistic sense as creative qualities. Mrs. Adobe, the film media teacher, acknowledged that her top students had a lot of creativity:

> Some kids shine more in Film Media 2 than others, some take the bare minimum that's required of them and go above and beyond and create the extra titles, and add in the extra effects, and a lot of that they end up doing on their own. A lot of that is trial and error, and "Hey, let's see if I can do this!" and their own creative juices that they'd like to have with it.

However, some students felt their imagination was stifled by the types of projects required. Oz complained, "Well we don't actually make like storyline movies. I think this could be fun. But we don't." Imagination was not only in creating media, but also in creating new code. Cecil was excited to show the interviewer a list of binary code he was

146

working on:

> Notebook paper to try out the first version of it. Here is the new one I am working on. [Cecil showed the interviewer a piece of paper covered with his handwritten list of binary numbers] That, I don't have a software to do it on. The largest thing I've done was probably five letters long and I can do it all by hand. Sort of slow and sort of lengthy but I like it because it converts.

Being able to brainstorm and create new ideas was important to these students.

The inventiveness side of creativity came out more in the programmer type students. For example, Cecil said:

> I have a lot of ideas for computers. I don't always have a chance to work them out. I like browsing Radio Shack computer. Last time it was actually yesterday. They have a bunch of little electronic pieces. I have a lot of ideas for working with those, but I doubt I'll get a chance to put them all together.

He was intrigued by what could be created with technology. But he was not limited by physical elements, and he also explored software. Cecil had fun breaking through code restrictions: "Basically the challenge, seeing what creative ways I could use things. I liked the idea that whatever they blocked, I'd probably find a way to use it." This enjoyment was shared by his friend, Dylan. To explain his non-traditional thinking, Dylan wrote, "I take advantage of every opening, I overthink things. I'm the kind of person who thinks about how to rob a bank just for the challenge." These kind of mental activities helped them have a creative perspective when approaching problem solving.

Finally, as mentioned in previous samples, there was an aesthetic appreciation present in many of the students interviewed. Four out of eight played a musical instrument or were artistic in other ways. Cecil was the most musically accomplished: "I play French horn, percussion, and piano." Euclid said he was involved in photography and also played the trumpet. Oz added that he enjoyed a lot of different activities, some of which were artistic: "I dabble in so much stuff. I inline skate, I like drawing, and then

147

there's the computer stuff, I also play drums." The four students in the media group enjoyed adding artistic flair to their movies above and beyond the requirements. Mrs. Java, another media teacher interviewed, recognized that these students had special abilities: "Kids who are really good when it comes to visualizing things, and that's not something that we measure in tests and everything." She felt that sometimes the education system did not emphasize these creative abilities as much as they should be.

Socioaffective.

The abilities of this category focus on communicating ideas to other people, social interaction, and trying to please others. The four students in the media group seemed at ease when talking about themselves and their projects. This could be because the sponsor, Mrs. Adobe, specifically chose students for this group who had high communication ability and could express themselves. She said a key component in her decision process was, "Their compatibility with the other students, because as much as it is independent it is also a lot of group work as well, and you're going to be spending the whole year with those five other kids. So they have to be able to get along really well too." However the other students not in the media group had varying degrees of communication skills. One student, Dylan, opened up and admitted that he did not feel confident in his abilities to talk to other people, "Without communication, it's pointless, and we get back to said shyness. It causes me to think almost anything would look weird to others in a social situation so... I suck at reality." He preferred on-line communication, and was happy with his connections on there, until they were taken away from him as a punishment: "That had been my support system, entirely. So I fell apart, that's why my sanity goes in and out nowadays." He admitted that he did have a group of friends to socialize with

148

outside of the virtual environment, but he still liked using a keyboard to talk to them instead of the phone.

In relation to interactions with the opposite sex, it was the teachers who observed the students' social abilities. The middle school media teacher, Mrs. Frink, noticed that, "They work better independently, we are trying to get them, trying to build more of their social skills. Try to pair them with the girls." But not all students in this sample were stereotypically awkward, and interactions varied by individual. The gifted teacher Mrs. Basic described the differences she observed:

> The boy, who wants to be a computer programmer, he's a little more what you would call, uh, well what is he… bookish, not very social…. Not really what you would call strong on the social skills, difficulty maintaining friendships. But the other girl who's not really into the computer side of things, she's more into the film making equipment, very social, very well liked, active in activities and stuff.

A revealing story was related by Mrs. Adobe, the media teacher, in relation to technology talent and social skills: "There's a student that works in here frequently. His name is Randy, and he's autistic, but he is fabulous with these programs. If he had the other necessary skills to be a Phoenix Films student, he would soar with the technology aspect of it." Here was a student with a disability that showed extraordinary talent with computers, but had poor social interactions. The other students also mentioned him and were in awe of his knowledge, despite his other deficiencies. Therefore social interactions in this sample ranged from isolated to very active.

Something interesting that came up in this sample was that only a few felt their skills were impressive or "magical" as other samples had mentioned. The mystique seemed to have worn off as more people became comfortable with computers. However there was still some spark there, as Dylan joked: "I've seen too much of the internet now, not like I get to use it ever... but then, I'm magic like that. No, actually there was a period when I did get to see the entire internet, all of it. All the internet, and don't ask questions!" Even if the magic had worn away, some participants still enjoyed knowing tricks that other students did not. For example, when Cecil was asked if people treated him differently because of his interest in computers, he responded:

> Not especially. Of course they will refer to me because I've got a graphing calculator. I am one of a few people who ask for a graphing calculator for a birthday. I was referred to today as that kid with graphing calculator with the artificial intelligence. I said I wished. That would be cool.

More often, the students interviewed talked about actually hiding their skills or being embarrassed about their technology interest. Euclid wrote, "Junior high I kinda tried to hide my knowledge. I didn't want people to think I was geek, so I didn't talk about it all." Some actively worked against displaying their abilities, like Amos: "In elementary, I was in the gifted program; mostly for math. Back then I thought I was a 'nerd' for being smarter than most kids in math, so I didn't stick with it, and now of course I look back and wish I would have." This change from the other samples is a concept that merits further examination.

Intrapersonal Catalysts

   Motivation.

Because the students in this sample were still in high school, motivation centered around school work and how it could come into conflict with their technology interests. Many of them had goals to learn as much as possible about technology and computers. Oz said, "Every time I try and do a project, I try and do something that I haven't done before, even if it's small, just to learn a little bit more about the program." But it had to be knowledge that interested them, or they could easily ignore it. As Euclid stated, "Well

150

see, I read a lot and I like to learn stuff. I don't read and I don't learn in school, but when it comes to stuff I enjoy I read articles online all the time." Amos's father commented, "I would say Amos wants to do the best he can at whatever he is doing."

Some were aware that their own goals often came into conflict with the academic goals set by their school. Dylan admitted, "I just never do the stuff that actually has a bearing on my future. To tell the truth, I'm just content to sit there and code, mess around, I don't need that much really." The gifted teacher, Mrs. Perl, saw this as an impediment to their school success.

> They spend a lot of time at home on the computers, and the parents complain that, especially with a few of the underachieving gifted kids, that it's hard to get them to do their homework because they're doing their computer stuff and they should be doing their homework.

However, most of the students were smart enough to make the system work for them. They could easily see what work was important or not, and what was needed to maintain a good profile in school. Euclid wrote, "In school I don't do homework in math.... I get like 90s or 85s on tests without doing homework." Most of them had good grades, but that was not a main motivator in their quest for technology knowledge.

The majority of participants clearly stated that they were intrinsically motivated to pursue their own technology interests. It could be hard to explain at times, as Euclid talked about, "I just had something inside me that wanted it, and friends and my dad kinda let me do whatever I wanted." That hunger or passion for technology verged on obsession with a few participants, like Dylan: "A teacher I once had said it was like a drug for me." But overall, the desire to learn about technology was positive, and they would take every opportunity to expand on their knowledge. Oz felt compelled to always tinker with his friends' computers:

I just really like keeping a well maintenanced computer. I'll go over to friends' houses and notice how slow their computers are. I can't help but run virus and spyware scans, and get rid of background applications that take up to much space. They are all grateful, but they also call me a giant nerd afterwards.

In relation to this quote, Oz's father observed: "It wasn't until junior high that he took a liking to HOW the computer worked and what was inside. Though even today, I wouldn't consider him a computer addict or guru. He knows a lot for a kid his age but it isn't his main focus in life." So even though he felt compelled to troubleshoot computers when he came across them, he also had a passion for other activities in life as well.

The more technology experiences they got, the easier it became to generalize their knowledge to different programs. Mrs. Java, the media teacher, was aware that these students could motivate themselves if she gave them technology-related assignments.

There are some kids who seem to have the intuition about, well ok, I can use this set of knowledge to springboard to this next program. We may not have the time to instruct them on, but they know the layout and everything, and they are motivated internally enough to go figure it out.

As graduation approached, some of the older students talked about the types of jobs they would like to do after college. There was some sense of external motivation there, but it was still fuzzy and ill-formed. Tex said, "I'm going to study electronic media production, and I hope to get into being somewhere behind the scenes at some sort of television station. Hopefully MTV (Music Television)." The types of jobs they talked about were areas that stimulated their interest, and that they would like to find out more about in the future. Euclid stated: "I've always thought it would be fun to build microchips. Dunno what that takes... but it would be cool." As they get older, the external rewards for their computer technology talent might become more apparent as they enter college and figure out what they want to do with their lives.

Volition.

With this sample, many of them talked about persistence and effort and not being afraid to explore software and hardware. The middle school media teacher, Mrs. Frink, noticed these qualities and noted, "They are fearless and I think it is more of a generational thing. They have grown up with them and that is their second nature thing." Johnson (1997) also observed this phenomenon in his research:

> You can see this aptitude already in the generation of kids raised on video games. There's a certain fearlessness they exhibit upon entering into a new information-space. Instead of reading the manual, they'll learn the parameters in a more improvisational, hands-on fashion. (p. 228)

Mrs. Perl, the gifted teacher, contrasted her students' persistence when it came to trial and error with her own generation:

> I think they spend so much time at the computer and they just aren't afraid to go in and figure out if something crashes like us older people who freak out and have to figure it costs thousands of dollars to get repaired. They go in and figure out how to do it. I mean there will be kids that can take things apart and put it back together. Like years ago when we didn't have computers, the kids that took the clocks and the watches apart and put them back, those are the computer kids that you see now.

Dylan demonstrated this tinkering inclination by talking about the computers he makes from spare parts: "I salvaged computers. Built new ones from the parts... learned Linux basics with that." When asked where he got the parts, he responded, "Dumpsters mostly, people put whole computers in dumpsters. I just grab them." Cecil also noticed this difference between technology talented individuals and other people: "I know some people are absolutely certain that they can't learn to use computers and I've noticed that some people are a lot faster at picking things up on computers than others." If people were too afraid to try different things on their computers, then they would not get as far as these students.

153

Related to persistence was the task of expending effort on a technology problem. These students had the capacity to work long hours, but they would be selective about their activities. Like Euclid wrote, "I'm kinda lazy, but it all depends. My priorities are big." Cecil encouraged other people to try and learn not to give up too easily, "I think that thinking you can use a computer helps, but also just the attitude some people take is important. If you'll give an honest effort to figure things out, and if you can pick things up fast, you'll probably be able to use computers pretty well." They described themselves as lazy in certain areas, but on a computer they could work non-stop. Yorick talked about how he spends all his free time creating movies at home on his computer, and that his vacation was already planned out: "I'm going to make like 50 movies this summer, there's just one really big one that I have to work on." Dylan related his own story about working long hours while talking about the qualities important for technology talent:

> It's just the right reasoning and math skills with enough curiosity and people that don't ask someone else what to do when it doesn't work, they actually figure it out for themselves. That's REALLY important. I mean, I've spent 12 hours straight on one simple problem before, just trying to get it to work until it worked, and in the real world, not all question have answers yet. Sometimes you have to be the first one.

The projects that they focus on can be so idiosyncratic that their parents do not see the point of them. Dylan's parents worried that his long hours were symptoms of Internet addiction, and so they reduced his computer access. The intrapersonal aspects of persistence and effort were present in most of the participants in this sample, and they were still trying to figure out how to balance technology with other areas of their lives.

Self-Management.

As with the other samples, the participants in this sample emphasized that they were self-taught in the area of technology. When asked during the focus group how he learned new programs, Jade said:

> I think like a lot centers on the self-teaching aspect, like you can only learn so much from a teacher that isn't thoroughly trained in all the programs and stuff. So you just have to learn yourself, and I think that if you teach yourself how to do it, you can go a long ways.

Even when they did have a class related to computers, they preferred to learn by themselves. Oz stated, "I was forced to take Keyboarding 1 in junior high, but other than that I really haven't taken it. I kind of like learning it by myself." In Cecil's case, the gifted program allowed him to teach himself using the textbook and test of the keyboarding class. He felt this was better, because he could accomplish more at his own pace. Cecil though this was an aspect important to his technology talent development: "Curiosity is important if you're actually interested in what's going on, you'll probably figure it out eventually." Many of the resources these students used to teach themselves were available on-line, and they would search for answers through their computers first before asking someone else for help. Oz mentioned that, "I did most of it on my own.… The internet was my best friend when it came to new stuff." Almost all of the participants specifically said they learned most of their technology skills by themselves.

There were still aspects of self-management that these high school students needed to work on, like scheduling and organization. The gifted teacher, Mrs. Perl, noted that, "They don't organize their time, in terms of how they use their computer and the time they spend on their computer, compared to some of the other things." The other gifted teacher, Mrs. Basic, also talked about keeping these students focused on their work. She recognized that her students were mainly self-taught, but she still tried to guide them and keep them headed in the right direction:

They just don't have time in their schedule, due to all their requirements and we don't have any wiggle room for some of these kids. They have to do it on their own. And when you do it on your own, you don't know if you're going down the right path, and you lose sight, and you're off track, so it's easy to get kind of distracted.

In Dylan's case, this idea of distraction and fractured attention could be related to aspects

of ADHD (Attention-Deficit/Hyperactivity Disorder). He admitted to having difficulty

paying attention in class, but he had no problem focusing on technology: "All I know is if

I can just use a computer for about 8+ hours straight I'll end up feeling better. It's the

only thing I know." However, many of the students in this sample talked about multi-

tasking and doing many different activities at once on the computer. They could have

multiple programs running at once and divide their attention between these tasks. Oz

wrote, "It's slightly embarrassing, but I spend 90% of my time at home on my

computer…My computer is my portal for music, research, videos, and my own films."

Even though they could do a variety of activities on the computer, some aspects of life

could be ignored. Mrs. Perl saw it as important to help her students disconnect from the

computer every once and awhile to become a little more well-rounded:

> Some of them just don't have a variety of experiences they spend so much time at the computers that they don't know some other things that might be happening out there in the real world. They are so in tuned to the computer stuff, but they know everything about all the computer companies. They're definitely focused on one thing.

Personality.

Adolescence is a major time for identity formation, so the personalities of these

students were not necessarily fully-formed by the time of the interviews (Schultz &

Delisle 2003). When asked if he participated in any social networking sites on-line,

Dylan responded, "I just try not to be mainstream, and anything like that is just... ew." He

enjoyed being a non-conformist and took pride in being an outsider at his school. Yorick dressed and acted a bit like an oddball, talking about his *Star Wars* fan movies and showed the interviewer his own clips on YouTube. This group of students had a quirky sense of humor, and took every opportunity during the interviews to tell jokes, related to technology or not. Oz said in his follow-up e-mail that he has, "A really big aspiration to be a stand up comic. My friends find me funny, a little crazy, but I think I'm pretty down to earth too." Mrs. Adobe, the media teacher, said she looked for a certain type of personality for the students she picked for the Phoenix Films:

> You have to make sure that they're responsible enough, because since it is an independent study, you have to be able to trust them with the equipment, to go out on their own, to know that they're going to represent your department well, that they're going to be polite, that they're not going to mess around.

Therefore any student interested in technology who might be withdrawn and anti-social was not someone she would pick for the independent study media group.

The temperament of this sample varied depending on the individuals interviewed. When asked to describe his personality, Euclid wrote, "Sporadic, crazy - I kinda yell random stuff really loud when I get upset…. I know I can get uptight but I try hard not to. I try really hard to see both sides of every situation, that's kinda important to me." During the group interview with the film media students, they seemed relaxed and friendly, and they talked about how being outgoing helped them complete their film projects around school. Cecil was the only one who was noticeably quiet and reserved, choosing to let his friend Dylan do most of the talking during their interview. Dylan was very animated, and said, "You shoulda seen me last year when I had internet withdrawal… My friends say I was really grumpy and stuff, and a bit psychotic." On the opposite end of the spectrum, Oz's mother said about her son: "He is slightly introverted, but is social. He prefers a small group of close friends over a large group. Oz thinks deeply about most things, is very sensitive to others and has a mature wisdom about him that seems beyond his years." There was a spectrum of personalities and temperaments in this sample, and they resisted being generalized or labeled.

Environmental Catalysts

　Family.

For some of the students in this sample, they grew up with parents who had careers related to technology and media. Amos wrote,

> My parents had a little head start in the 'technology' world, but it wasn't long before I was showing them how things worked. My Dad is a graphic designer and does a lot with Photoshop and Illustrator, so we both share tips and ideas; my mom pretty much only checks e-mail, websites and plays the occasional computer game.

He was able to learn along side his father and share ideas. Euclid's father was more of a tinkerer, but, "He wanted to be an engineer. His counselor told him his math grades weren't good enough, so he did business and worked for a government operation. But he's building an airplane now." This influenced Euclid's interest in robotics and technology. Cecil's father, an architect, was also technically minded, but he often deferred to the knowledge of his son when it came to the computer: "My dad gets worried sometimes that I'm going to mess something up but every time my mom has something she needs to do on a computer, she writes it down for me and has me do it instead." The educational background and career choices indirectly influenced the students.

In a more direct influence, the parents were responsible for the acquisition of computer technology in the household. Tex wrote, "My parents bought me and my brothers a computer game called *Spellicopter*. That was the first time I got on a computer." Computers and educational software seemed to be what parents purchased to increase their children's skills and knowledge. Amos remembered one such program,

"Before too long I was playing *Mario Teaches Typing* which really did help my typing skills. I never thought that computers would evolve so fast." Sometimes it was other family members who bought computers for them, as in Oz's case: "I received a computer from my grandfather, and kind of took it upon myself to learn more about it. I went pretty deep too." By acquiring a computer for their children, they hoped to positively affect their learning and encouraged them to use the technology. Jade stated: "My parents were very supportive of my technological interest… they paid for that first computer, and while they don't necessarily pay for my current computer endeavors, they're still very supportive."

Another facet of parenting that was brought up during interviews was the rules established about computer use. Amos's father wrote, "We did set time limits on how long our children could use the computer." Most of the participants started out with limits, but they were usually flexible. Cecil remembered, "When we first got a family computer, the rule was if I was going to play on it, I had to play a half hour of educational games first. But after a year or so they just stopped caring." Also, as they grew up and became more independent, guidelines changed to suit their development. Oz's mother stated, "We had fairly strict rules about the amount of time he was allowed to play on the computer…. We did not allow him to have a game system like Nintendo until he was nearly junior high age." But Oz was able to make choices on his own about technology once he was in high school:

> When I was younger they really set limits on the computer. However, this past December, I bought my own computer with my own money, so they don't mind as much. As long as I'm keeping high grades and I'm in a good mood, they let me do as I please.

Not only were there rules about time limits and content, but punishments could also be linked to the computer. Dylan described a particularly hard time for him at home when he was punished: "My mom just decided to take all my computers away. And then they just kept it going because I was very difficult about it. Let's just say that this is how I learned to assemble a computer in the dark." There could be a power struggle over the computers, and some of the students would find ways around the restrictions.

School.

Besides having computers at home, there were also technology opportunities at the school. This particular school district has recently pushed technology improvements as one of its most important goals, and the comments by the students reflect this initiative. Jade remembered an early experience before the district reform: "I recall being exposed to technology around 2nd grade when I started learning the basics of Microsoft Office in computer classes at school." It was not until the high school level that they had the opportunity to take a variety of technology classes. Amos outlined his current course load: "Many of my classes are primarily on the computer: electronic media, film, yearbook, Phoenix Films, and next year the architecture will be all on the computer for the first time." There were also extracurricular groups at the high school related to computing. Euclid enjoyed being part of a team, "The robotics team was sponsored by the gifted teacher, and we get along well. I was kind of a favorite of his on the team." There were also computer gaming clubs established at both high schools in the district. For some levels of technology, the school was the only place with the resources. During the group interview with the media students, they were asked the question, "Would you have been involved in these media projects if you didn't have this technology available at

160

your school?" Tex replied, "Oh definitely not. I definitely would not be involved in it... I don't have these programs at home, I don't have a computer that could run them."

The positive influence of school was also balanced by negative experiences. There were those students who expressed a general hatred for school, like Yorick: "I don't do anything with the school. School makes me mad... Because it makes me come here for 8 hours of the day." He preferred making movies at home and not getting involved with formal education. Other students related specific incidents where they clashed with teachers, like Euclid:

> I took computer programming sophomore year. I had an A for the first quarter but I asked too many questions and questioned her tests and stuff, so she got upset and told me she'd get me expelled. I asked her some question on a test and she said 'I'm the teacher, I'm right and you're wrong.' So I said, 'That's bullsh*t,' and she FREAKED out, and started yelling at me. So then I got a C somehow.

Mostly they just wanted more freedom within the school environment to experiment with programs. In middle school, Cecil said:

> If I had access to a computer, just during any free time, more access to computers, I'd probably be further along. Because often I can't find... that's why I figure out things just by trying. If I can't find any good book, which I can't usually. I don't get much of chance to find books on it.

When the interviewer asked Mrs. Basic, the gifted teacher, if the students had free time on the computers, she said: "They don't, because if they are just 'messing around' that would imply that they were off task, and that's a big no-no, you can't have that. So there's not a lot of flexibility." Some of these students expressed frustration that what they thought of as self-directed learning was not considered valuable by some of the teachers.

The teachers had an effect on the development of these students, positively and negatively. Dylan was happy to get to high school and find a teacher he could connect with: "My current computer teacher is the only person who has ever been anywhere near my level." This teacher was able to give him guidance and help him enter a programming contest, where he won in his division. Many teachers did not have extraordinary technical knowledge, and so they looked for alternative ways to support their students. Mrs. Adobe the media teacher was aware of her role in the classroom:

> Once they're in Phoenix Films, I don't really teach them anything else – they teach me. But, I am the one that sets up deadlines that they have, and places that they need to go, and people that they need to talk to, but as far as the projects are concerned, I'm more of just a warm body, a facilitator.

She understood that a collaborative team environment worked best for the students in her class. However if a student was interested in technology but did not have room in his or her schedule for a class, they could be denied access to areas of the school. Mrs. Basic, the gifted teacher, described the frustration some of her students faced when trying to work within the school boundaries:

> We actually had some complaints from some of the kids that they didn't have access to certain resources because they weren't enrolled in a class. Or they wanted to do something in terms of a competition, they wanted to use some resources to compete in something, but they weren't allowed access to the resources here because they weren't enrolled in the class and therefore the teacher was not going to just let them loose on her equipment.

She felt that the school needed to continue to evolve to try and meet the needs of all of the students in the building, especially those in the gifted program. Oz's mother wrote that her son, "Had no interest in gifted programs for math and science in junior high, but I think he would have participated if there had been a technology component to the program." If more changes were made in the required curriculum at the middle and secondary level, then it would be possible for more students to expand their knowledge and skills with computers and technology.

162

Peers.

In this sample, the students seemed more comfortable with their peers in relation to technology. Amos stated, "With the growing trends in technology, knowing a lot about computers isn't very unusual among highschoolers." They could share their interest with computers with friends and not feel like they were different. There was a varying degree of peer involvement in their technological development, from strong to slight.

For some, there was a direct influence from friends, like Euclid, "Third grade my best friend was into computers, and we were competitive so I had to have one too, but better. I got better than him quickly, and knew more." In other participants, they served a particular role in their social group. Amos wrote: "I would say I am considered a computer tech among most of my friends. I am usually one of the first ones asked when someone has a problem or they are troubleshooting something." Also proximity to friends was mentioned as a factor in Euclid's case. He felt that his rural community impacted his communication with his peers. The computer became his connection to his friends, as Euclid explained, "I had friends at school, but that was 20 minutes into town and my parents didn't have time to drive me. So AIM (AOL Instant Messenger) was easier than the phone."

The teachers also commented on the importance of peers. Mrs. Perl, the gifted teacher, observed, "They tend to hang with the other computer kids and some of them that I see don't necessarily have a whole bunch of friends, they're not involved in a lot of other things at [high school] they're so into their computer stuff." They are drawn to peers with similar interests, but it also varied by individual. The media teacher, Mrs. Java, discussed the peer interactions she observed in her classroom:

It depends. I think if they feel like they have a peer that they can relate to, you know like some of those visual kids I was talking about, they play off each other. The one I was thinking of, the whole analyzing film and everything, he was a loner. He was perfectly fine to not need to talk to anybody. He would if forced to, or encouraged to, but he wasn't drawn toward that.

Having a friend who was also into technology could be beneficial, but it was not a necessity for some students.

When a close relationship with a technology partner is formed, the dyad seems to compliment each other. In this sample, there was a clear dyad between Cecil and Dylan. Both of them had originally been interviewed in middle school, where they worked on many computer projects together. Cecil animatedly talked about his working relationship with Dylan:

Then at the library once with my friend Dylan he and I wanted to know how to make a bat file. So I found that at the library and that was exciting, and I always wanted to learn Java.… I got a book on it and Dylan also got a book on it.

They enjoyed working on ideas together in a variety of formats, and Dylan said, "I've got a bunch of notes over there with stuff that has to do with computers. So sometimes I just do it all on paper and sometimes we actually do it on computers. It differs really." It was also apparent that they had complimentary interests when it came to the different aspects of computers, as Cecil said, "I don't know the gigabyte/megabyte… That's Dylan's thing." Also in middle school they self-identified as the top of their class with technology. Cecil said, "In school when somebody is trying to do something on the computer and it is not working out, they call me or Dylan." Once they reached high school, their partnership moved more on-line, but they still chatted with each other. Cecil stated that when he programs now, that, "Generally I like to be physically alone, but working with Dylan over chat or email." There was less time for them to get together, but

164

they still collaborated when they had a chance. Dylan wrote, "I can't do much right now, but we work together sometimes... But it's just too hard to get somewhere I can put in the time." A dyad such as this can fuel technology projects and further exploration.

Some students preferred to be in a homogenous group of computer enthusiasts, but the majority talked about being part of a diverse social circle. This reflected their opinion that they have diverse interests and are not necessarily labeled as only a techie with their friends.

Talent

The different types of technological activities this sample enjoyed varied by individual. Some of the things they liked to do on the computer included: programming, hacking Windows and other operating systems, playing games, downloading music and videos, instant messaging, posting on message boards, creating films, researching for homework, and posting on social networking sites. They liked to try many different activities and would stick with the ones that they enjoyed the most.

The most frequent activity mentioned by the participants in this group was multimedia production. Seven out of eight interviewed mentioned creating movies and videos at some point on their computers. This could have been influenced by the fact that four out of eight interviewed were current members of the film media group, Phoenix Films. When Jade was interviewed on-line, he wrote that: "I'm doing a few peoples' graduation videos right now, actually." Amos, Tex, and Oz all aspired to work in media production and enjoyed creating movies at school and at home. However, even the students not in the film media group enjoyed creating movies, especially Yorick. Yorick showed the interviewer several movies he had made on YouTube and enjoyed explaining

165

all of the technical aspects related to filming and adding the special effects. During the group interview he even got into an argument with Dylan about the best way to make a person's arm come off and fly across the screen. Dylan also talked about making movies during his free time:

> I think I'm going to make like two movies this summer, somehow. My brother is going to get forced into making one because they bought him that expensive camera so he's going to use it. And I'm going to borrow that to make my friend's movie, which I'm going to end up having to edit.

In middle school, Cecil talked about creating humorous movies using pictures and music, but that he no longer had time to play with that program in high school. Although most of the students only had access to the film editing software at school, they hoped that the programs would become more affordable in the future so they could have more time to work on their projects.

Three out of eight of the participants in this sample labeled themselves as programmers and talked about hacking and coding. Euclid wrote, "I was a power user, and I knew every nook and cranny of Windows XP: command line, networking, system config, shoot from the run command." It was fun for him to explore Windows and find parts of it hidden to the average user. Cecil was the same way, and in middle school he talked about the joy of finding out one of the functions in Windows:

> Then I found Command Prompt. And I started learning that maybe a year ago. And how I learned the coding was by using the help command. 'H-E-L-P, Enter' and it listed a bunch of the commands and I didn't understand anything. So I think the first day I learned how to actually add it so I can open various documents and it was really exciting and it did not go very far until I learned help command. And then I started learning more and almost everything I learned from then on I learned to do through the Help command.

Once he discovered this ability, he created programs to run and shared his ideas with his computing friends. In particular, he worked with Dylan, who labeled himself as a hacker.

166

He said that he used to moderate a message board devoted to software hacking, but that

he did not participate as much anymore now that his computer access was limited. When

I asked him about what he did in his old hacking days, he said: "Just messed around a lot,

that's how I learn. Oh, and never ask a hacker, 'How do I hack?' Worst question in the

freaking universe." All three mentioned that they mainly explored programming because

it was challenging and they enjoyed problem solving.

All of the students in this group used technology for entertainment and

communication in some form. In relation to playing video games, some were more avid

players than others. For example, Amos considered himself a frequent gamer: "A couple

of times we have connected four Xbox 360s together in our own network and played 4 on

4, which is a lot of fun, especially if you're like me and win a lot." During the group

interview, Dylan played with a game on his calculator but admitted that he did not

consider himself a big game player. For Cecil, he was more fascinated in finding out how

the games worked than playing them:

> I am not as much into computer games as some other computer people are. I've got
> interested in text adventures recently because I can make them. I know how to make
> them with MS DOS. But I found that most text adventures online are not in MSDOS
> coding. They are mostly in Java and Apple most of the time and I was disappointed
> by that because I can't look at the coding.

They all occasionally played computer or console games, but saw it as more of a social

activity with friends and did not take up too much of their time. Besides entertainment,

many of them used social networking sites to stay connected to friends on-line. Oz

responded, "Well, being a teenager, I've used Xanga, MySpace, and Facebook before."

Most of them had a personal account on one or more social websites, but the three

programmers did not like these networks. Yorick and Cecil both disliked putting personal

167

information about themselves on the web, and when Dylan was asked about MySpace he wrote, "It promotes idiocy and narcissism." Therefore the participants used technology to communicate with their friends, but some were more deeply involved in socializing on-line than others.

Predictions for the Future

This sample was very positive about the continued integration of technology into society in the future. All of the members of Phoenix Films were excited about the possibilities for media production and editing software. They desired to continue exploring the film media programs, and predicted that the capabilities would continue to improve. Tex said: "The programs, they do so much that I don't even think the designers of the program know what exactly it can do. So, there's no limit." In relation to media production, Mrs. Java talked about the journalism class adapting to the way news has evolved:

> Some areas of our broadcast class and newspaper class that we've tried to work on is converging media, since that's going on in the real world, and so we should be trying to do it on whatever scale we can, even at the high school level. Because once people go to college, try to get a job, in the industry they're no longer just in print journalism, or just video journalism… They're expected to be able to do different areas all at once.

The students and the teachers recognized that technology knowledge would be important for their future careers, regardless of their current skill level.

Although there were some strange statements made like, "Robots take over the world," as Yorick wrote, most of the students had a realistic perception of how technology would affect society. Cecil described the functional aspect of technology in the future:

Random household appliances will continue to get more and more complicated computers, as those get cheaper and computers will continue to improve in speed and possibly quality, but the internet will become more and more integrated into basic programs, like it is with media players now.

Increased integration of communication was mentioned by several participants, as they talked about improvements they would like to see in the near future. Dylan had his own take on how the internet would evolve:

I recently had some cool theories about neural interfacing, but in general I think that things are gonna just keep getting smaller and more personal. Eventually, the internet might start blurring even more into reality until we don't even THINK about it, it's just there. We won't call it going on line, using the internet. It will just begin to blur until possibly there may be NO distinction.

They had an optimistic outlook for new and improved technology, and they looked forward to expanding their skills and knowledge in this area.

Summary

The participants in this sample expressed the belief that having logic and excellent problem solving skills were important in the area of intellectual abilities. Half of them recognized that having a high level of math abilities was helpful in relation to programming, but it was not a requirement to have computer technology talent. They emphasized how important creativity was to solving problems, and they displayed imagination and inventiveness in their work with computers. The film media group were dedicated to creating multimedia projects, and worked year round to produce films and video compilations for the others students and teachers in the school. There was a clear difference between the socioaffective abilities of the Phoenix Films group and the other students, as it was a requirement of the independent study class that they work together and be polite and friendly. The other computer students were more awkward and had fewer social skills, an area where the gifted teachers thought they could improve upon. In

169

terms of motivation, these students were fueled by their passion for computers and technology, which sometimes came into conflict with their academic achievement. They were not afraid to try new things while exploring their areas of interest, and they put a lot of effort into working on their own technology projects. Being self-taught was important to them, and many of them sought out a variety of ways to use their knowledge and skills. I observed that the majority of these participants had pleasant and easy-going personalities, and the ones who were oddballs enjoyed sharing their quirky humor with me. Overall they had a normal range of personality traits and temperament, as much as could be expected from a group of adolescent males.

As with Sample 3, families acquired computers and educational software for these participants and introduced them to technology at a young age. The difference in this sample is that the parents specifically mentioned how they set limits on computer and internet use for their children. Some parents directly helped these students learn some skills, but mostly they allowed them to explore on their own. Many of these participants even had their own personal computers separate from the family one. Peers continued to be a strong influence in this sample, and it was through their friends that they learned about new programs and shared their knowledge and ideas. Many of them connected to friends on-line and used the resources available in internet communities to learn even more about technology. Within the media department at their school, they had just upgraded to new software and computers to help these classes create high tech film projects. These individuals were happy with the resources available, but the teachers pointed out that only certain students were allowed access to this area. Besides that, other formal education classes were disappointing to most students, and they school was not

170

considered an influence on their computer technology learning. They expressed their talents in a variety of ways, and a majority of the group enjoyed video and multimedia production. Overall, they were positive about the future of technology, and looked forward to what new developments would take place. They were interested in pursuing careers where they could use their CTT, but many had not yet decided on what that would be.

CHAPTER 5

Discussion

The purpose of this study was to describe the distinguishing intellectual and personal qualities of adults and adolescents who demonstrated computer technology talent (CTT). Four distinct samples were analyzed and the overarching trends were analyzed to help trace the developmental path of CTT in the lives of the participants. Overall, there were similar trends and patterns across the samples as well as some notable differences. The emergent results were most similar to Gagné's (2003) Differentiated Model of Giftedness and Talent (DMGT), and this model was used as a conceptual framework for the conclusions of this study after the initial analysis was conducted. How this study relates to Feldman's (1994) co-incidence theory and universal-to-unique continuum is also discussed. The limitations and implications of this study are examined, as well as suggestions for future research in the area of gifted child studies.

Conclusions

The research question of this study asked about the cognitive and affective qualities and life events that mark the development of CTT. I believe that these characteristics were found within the qualitative evidence collected from the four samples. The trends are divided below according to Gagné's DMGT categories: (1) natural abilities, (2) intrapersonal catalysts, (3) environmental catalysts, and (4) talent activities. All of these factors in Gagné's original DMGT were explored except for *chance*. Feldman and Goldsmith's (1986) term *co-incidence* was used instead of *chance* to account for that essential part of talent development.

Natural Abilities.

Under the heading of natural abilities, there were three concepts examined: intellectual, creative, and socioaffective abilities. Gagné (2003) emphasized these as areas of giftedness where a person could show potential at a young age. For this study, I used my own interpretation of the term *socioaffective* to represent the ability to interact with other people and successfully communicate one's ideas. This allowed for a more malleable description than the one originally presented in Gagné's model. The general trends for natural abilities between the samples are presented in Table 5.

The intellectual abilities mentioned in this study stayed fairly consistent from sample to sample. All of the samples emphasized logical thinking and problem solving as important to working with computers and technology. Less emphasis was placed on memory and processing speed by the later samples. Sample 1 started out with a heavy emphasis on mathematics, and most of the programmers in this group considered it a requirement for computer technology talent. However, there were others in that sample who thought that having a math interest was only one path a person could take in computing. In the later samples, some individuals continued to be math-oriented, but the groups did not considered it necessary to understand how to interface with computer programs. All of the samples recognized that intellectual abilities were necessary as a starting point, but they placed more value on creativity as a determiner of technology talent.

On the whole, the creative abilities discussed were similar for all of the samples. Having imagination and creating thinking was considered important to all of the participants in the different groups. The qualities of inventiveness and innovation were

173

Table 5.

Trends for Natural Abilities

	Historical	Snapshot	Longitudinal	Contemporary
Intellectual Abilities	• Intelligence • Logical thinking • Problem solving • Memory capacity • Fast mental processing • Math-oriented	• Intelligence • Logical thinking • Problem solving • Depth of knowledge related to technology	• Intelligence • Logical thinking • Problem solving • Some math-oriented	• Intelligence • Logical thinking • Problem solving • Some math-oriented
Creative Abilities	• Imagination • Innovativeness • Aesthetics found in computer code	• Imagination • Creative media projects • Some artistic and musical	• Imagination • Creative media projects • Some artistic and musical	• Imagination • Creative media projects • Artistic and musical
Socioaffective Abilities	• Communication facilitated by computers • Self-described awkward, virgins and nerds • Used computer "magic" to dazzle people	• Communication facilitated by computers • Somewhat awkward, virgins and nerds • Used some computer "magic" to dazzle people	• Communication facilitated by computers • Less awkward, better social interactions • Need better group working skills	• Communication facilitated by computers • Less awkward, better social interactions

most important to the historical sample because the programmers were still in the process of creating so many new products. It was important to Sample 2 that they knew what the rules in computing were so they could figure out creative ways to break them. The third and fourth samples were more interested in using the computer software creatively to produce media projects. In relation to aesthetics, Sample 1 indicated that they saw beauty in a piece of code and enjoyed seeing it written cleanly and efficiently. Later groups placed emphasis on creating beautiful music and art with the technology available to them. Samples 3 and 4 were especially interested in capturing beauty with video and graphical art programs. Their choice of creative medium also reflected the change in the technology capabilities over time. All of the participants talked about creativity as being more important than most other skills or abilities.

The way in which the participants described socioaffective abilities related to CTT varied across samples. All of the groups mentioned using computers to facilitate communication with other people. In Sample 1, the participants talked about how they communicated with people through their software and the products they created. The later samples used the internet to connect to people and talk to other computer users around the world. The historical sample talked about how most computer programmers fell under the stereotype of not fitting in and being socially awkward. Several of them said that there was just not enough time in the day for them to do all the coding they wanted and have a social life as well. As computers became more mainstream and accepted in society, so to did the types of people who were interested in them. There will always be some socially awkward "computer nerds" at every age, but there were more computer techies in Samples 3 and 4 who self-identified as being popular and well-liked and did not consider themselves to be "geeks." It is possible that the stereotypically low

social skills associated with CTT will continue to change as technology becomes even

ore integrated into normal society.

Intrapersonal Catalysts.

Building on top of natural abilities, Gagné (2003) theorized that the intrapersonal

qualities of an individual had a large effect on his or her talent development. The

different intrapersonal catalysts analyzed for this study were motivation, volition, self-

management, and personality. The trends for intrapersonal catalysts are presented in

Table 6.

The degree of motivation mentioned by the participants was different for each

sample. The interviews were conducted with four different age groups, and therefore the

priorities of the president of a computer company were assumed to be different from that

of a high school student. Sample 1 talked about being highly motivated and goal-

oriented, and their focus was creating products and running a successful business. Many

participants in this historical sample mentioned having peak experiences that drove them

internally as well. Other groups also mentioned being motivated to work with technology

for their own enjoyment, but most of them did not want to push themselves past their

comfort level to achieve higher goals. Sample 2 specifically talked about finding ways to

use technology to make things easier for themselves. They also warned against younger

generations getting into programming just for the money. Sample 3 realized they had

mostly slacked in high school and advised the next generation to try and do productive

things that could be put on a college application or a job resume. And Sample 4 said they

were motivated when it came to working with computers, but not as much when it came

Table 6.

Trends for Intrapersonal Catalysts

	Historical	Snapshot	Longitudinal	Contemporary
Motivation	• High motivation • Goal-oriented • Internal - Peak experiences • External - Successful businesses	• Mostly low motivation • Internal - Passion, satisfaction • External - Personal rewards, do not do it for the money	• Moderate motivation • Internal - Enjoy learning new skills, satisfaction • External - Getting into college, having a good job resume	• Moderate motivation • Internal - Passion, some personal goals conflict with school achievement • External - Desire for careers in technology
Volition	• Fearless with computers • Persistent • Put tremendous effort into work projects • Long hours / late nights	• Fearless with computers • Persistent • Put effort into own projects • Long hours / late nights	• Fearless with computers • Put effort into own projects • Long hours / late nights	• Fearless with computers • Put effort into own projects • Long hours / late nights
Self-Management	• Self-taught • Encourages curiosity • Take the initiative • Concentrate on one important task at a time	• Self-taught • Encourages curiosity • Not take initiative • Concentrate on multiple tasks	• Self-taught • Encourages curiosity • Some take initiative • Distribute concentration between multiple tasks	• Self-taught • Encourages curiosity • Some not take initiative • Distribute concentration between multiple tasks
Personality	• Mostly oddballs • Non-conformists • Quirky sense of humor • Easy-going to frustrated	• Mostly oddballs • Non-conformists • Quirky sense of humor • Easy-going to frustrated	• A few oddballs • Normal range • Quirky sense of humor • Easy-going to frustrated	• A few oddballs • Normal range • Quirky sense of humor • Easy-going to frustrated

to academic achievement. All of the groups were motivated to work on their own technology projects, but the degree to which they applied that energy to other areas varied by individuals.

The qualities of volition mentioned by each sample were very similar from group to group. The most frequently mentioned aspect was that they were not afraid to explore computers and technology. The original pioneers of technology in the historical sample were risk-takers, and their efforts paid off. This fearlessness was emphasized even more in the later samples. The teachers in Sample 4 in particular saw this as a quality that distinguished these CTT students from older adults. When these individuals worked on programs and ran up against bugs and errors, they persevered through the mistakes until they got the problem correct. There was a tenacity that prevented them from giving up too easily if it was a project they cared deeply about. Many participants from all of the samples mentioned working long hours and late nights when they were engrossed in complex computer problems.

In relation to self-management, some aspects stayed the same between groups, while others changed from sample to sample. Some participants mentioned taking the initiative as an important CTT quality. All of the samples stressed the importance of curiosity and taking it upon one's self to learn about technology. That desire to learn more and take in as much information as possible could be found in each group. If it was something that interested them, they would take the lead and start exploring. But samples 3 and 4 showed more hesitation if the project originated from an outside source. The importance of concentration decreased from the first sample to the later ones. The historical sample talked about holding formulas and code in their head for long periods of time and using immense amounts of concentration. It was a necessity of their job because the software originated in their minds. But as the interface became easier and more programs became available, the different generations began to multi-task more and more. They no longer focused on just one aspect but enjoyed diving into diverse aspects. They had the ability to concentrate for long periods of time on technology projects that interested them, but preferred to do many different parts of it at once instead of focusing on step by step procedures. This was most apparent when communicating with Sample 4 where they described doing several different activities at once, even while participating in their on-line interview with the researcher.

The personality traits mentioned by the different samples varied somewhat from group to group. The first sample had a lot of individuals who could be labeled visionaries and oddballs. They expressed characteristics that were considered very different from normal personalities. The second sample also had a fair amount of individuals who showed quirky and non-conformist personalities, and they were very anti-establishment. But the participants in samples 3 and 4 had a more normal range of personalities and blended into the school environment better. There were some individuals that stood out, but it would be hard for a teacher to identify who was a computer geek in these groups based solely on personality. All of the samples made jokes and described experiences that reflected their quirky sense of humor. The temperaments of the groups seemed to have a normal range from calm to easily frustrated, depending on the individual. Frustrations were mentioned most frequently in Sample 1 and Sample 2 because they had work environment stress to deal with and a personal investment in the success of their products. The younger samples 3 and 4 seemed more easy-going, but this could have been because they did not really feel challenged yet by the work required of them in their school environment.

Environmental Catalysts.

The external factors present in the lives of these participants impacted them positively or negatively, and had a direct influence on the path of their talent development. The specific environmental influences examined in this study were family, school, peers, and the media. The trends for environmental catalysts are presented in Table 7.

The degree to which family was cited as a strong influence changed across samples. While growing up, the historical sample did not have personal computers at home. Some had family members in engineering related jobs and had access to mainframe computers, if they were lucky. Their families supported their learning, but few of these individuals learned technology skills directly from their parents. As time progressed for each subsequent sample, the family became more important, especially when it came to purchasing the home computer. All of the students in Sample 4 had computers at home, and the majority of them had their own personal computer to use. The knowledge and skills of the parents also increased over time, and they passed what they could onto their children. Although there were still some parents who could be considered technophobic, they did not significantly hamper the learning of their children. An issue brought up by the parents in Sample 4 was a concern about internet addiction and a desire for them to be well-rounded and not focus solely on the computer. It was most important to all the samples that their families were supportive of their educational pursuits.

Table 7.

Trends for Environmental Catalysts

	Historical	Snapshot	Longitudinal	Contemporary
Family	• Some tinkerers • Encouraged learning • A few family members were gatekeepers to technology	• Some tinkerers • Encouraged learning • Acquirers of technology • A few direct teaching	• Some tinkerers • Encouraged learning • Acquirers of technology • Some direct teaching	• Some tinkerers • Encouraged learning • Acquirers of technology • Some direct teaching • Parents set limits on computer use
School	• Inadequate resources • Some math classes important • Universities had access to computers	• Inadequate resources • Unhelpful formal education • Allowed some access to school networks	• Inadequate resources • Computer programming club important • College courses finally challenging	• Adequate resources, but limited access • Computer media classes important
Peers	• Strong influence • Hear about computers through friends • Group dynamic while working essential	• Strong influence • Work on computers with friends • Connect to friends over the internet	• Strong influence • Work on computers and projects with friends • Connect to friends over the internet	• Strong influence • Work on technology projects with friends • Connect to friends over the internet
Media	• Changed the concept of computer programmer • Magazines presented business competition	• Some strong influence • Movies and TV shows centered on computers • Video games influence	• Not mentioned as an influence • Play some video games	• Not mentioned as an influence • Play some video games

The formal education options offered to these participants by the school system changed over time. There were improvements, but they were not fast enough to keep up with the needs of most individuals interviewed. As time passed from the historical sample, the school system attempted to increase their resources for technology education. Despite the efforts of schools, for most of the participants in samples 2 and 3 school always seemed to be one step behind. There were a few stellar examples of individual teachers that stood out, but mostly the education system was a disappointment. It was only with Sample 4 that the students in the independent study group said they were happy with the resources in the computer media department. However, they were quick to point out that this was a development that had only occurred in the last year due to the effort of the principal to upgrade their computers and software. The university experiences of the longitudinal sample proved to be more suited to their specific intellectual needs, and most participants suggested that CTT students should have access to college courses at a younger level.

There was a strong influence by peers mentioned by all of the samples. As self-directed as most of these participants were, peers played a big part in their computer technology development. In each sample there was either a best friend or group of friends with whom each person could share their love of technology. In the historical sample, intense working relationships were formed under the pressure of the original Silicon Valley businesses. The excitement of sharing this new technology with friends was one of the reasons they continued to explore new computer programs. Some of the participants in samples 1 and 2 mentioned that they initially got interested in computers at the urging of a friend. However, those stories decreased when it came to interviewing

Sample 3 and Sample 4. There was less of a specific moment of "getting into" computers as just always having one available and talking about them with friends. All of the samples recognized the importance of collaboration, and they especially preferred working with peers who had the same level of knowledge and interest as them.

The media played a unique role in the lives of samples 1 and 2 in this study. There was some mention of media influence in the historical sample, but more often than not the individuals being interviewed were actually the ones making headlines with their work. The participants in Sample 2 were the ones who most frequently mentioned the media as influencing their interest in technology. They clearly remembered movies, television shows, and video games as playing a large part in getting them interested in computers. Samples 3 and 4 did not mention that media was an influence in their lives, except for talking a little bit about the different video games they played. My conclusion from this is that in the 1980s and early 1990s computers were still new and special and the subject of many movies and TV shows. Then in the late 1990s and 2000s, the computer was no longer seen as brand new, and had become more of an everyday household item. Even if media had influenced the development of the later samples, it was on an unconscious level and not something they mentioned in their interviews.

Talent Activities.

The way in which each sample demonstrated their computer technology talent differed from group to group. The trends for talent activities are presented in Table 8. The talent area of the first sample was almost exclusively computer programming and hardware design. This result could have been because Lammers (1989) specifically chose programmers as the subjects for her initial interviews. Sample 1 also expressed their

Table 8.

Trends for Talent Activities

	Historical	Snapshot	Longitudinal	Contemporary
Talent Activities	• Programming • Software development • Hardware • A few create media • "Entrepreneurial giftedness"	• Programming • Software development • Hacking • Hardware • Networking • Communication • Troubleshooting • Some create media	• Some programming • Some hacking • Some hardware • Communication • Troubleshooting • Webpage design • Create multimedia • Diverse interests	• Some programming • Some hacking • Some hardware • Communication • Troubleshooting • Create multimedia, emphasis on videos • Diverse interests

talents related to *entrepreneurial giftedness* (Shavinina, in press). Having a talent for business requires that individuals have such qualities as risk-taking, being self starters, excellent organization skills, knowing how to market their products, and bringing together the best people to work on a team. It is not known how much of the skills and talents expressed by the historical sample were based on CTT or on entrepreneurial giftedness. This confounding variable is something that I recommend should be studied further in future research.

As data from the different samples were collected, the areas where they demonstrated CTT expanded. This was to be expected because the technology available to the later samples was quite different from what was accessible to the historical sample. Sample 2 consisted of current technology workers, and so the talent activities they discussed were based on their job experience as well as their hobbies at home. Each sample increased its ability to multi-task and create different media using the most recent computer technology. Digital communication was a big part of most participants' lives, and Sample 3 and Sample 4 stated that they enjoyed exploring every possible activity available to them on the computer. Some of them still continued to program and build computers, but not all of them. For the contemporary sample it was clear that their preferred talent activity was creating videos and editing films using computer technology. Whether they would end up carrying those talents into their future careers remains to be seen.

Co-incidence Theory

The four factors of the co-incidence theory discussed by Morelock and Feldman (2003) have all come together at the end of the 20th century to affect the development of

183

CTT: (1) The individual's life span, (2) the development of the field or domain, (3)

historical and cultural trends, and (4) evolutionary time (p. 462). The majority of this

study focused on the individual life span of the participants, but these results also relate to

the other aspects of the co-incidence theory. Specific to the development of the computer

science field, it is clear that the variety of technological activities available to individuals

has continued to expand over the years. People are no longer limited by a few choices of

software and hardware, and the area of creative multimedia software has seen tremendous

growth. As advancements are made within the domain, the knowledge and skills that

CTT children have will also grow to reflect those changes.

The interview data associated with each sample's predictions for the future also

related directly to Feldman and Goldsmith's (1986) second factor concerning the

development of the field. The historical sample had many positive predictions for the

future of society, and they were determined to be the ones to cause those innovative

changes. However, Sample 2 had a more negative perspective, and stated that they had

become disillusioned with what they originally had been promised about technology.

Samples 3 and 4 had more positive things to say about the future, and both groups

mentioned that they were interested in learning more in college and working in a

technology-related field. Their positive responses also reflected the optimism of youth,

and they had yet to experience the realities of the computing industry. These future

predictions show that these CTT individuals were interested in watching technology

grow, and most of them hoped to be the ones to make those developments happen.

It is too soon to predict the historical and cultural impact of this concept of

computer technology talent, but innovations continue to be made by these individuals that affect the wants, needs, and values of society. The next generation of children will be using increasingly advanced technology and imagining new possibilities for the future that could permanently change the way people live. Continued longitudinal studies could draw a better picture of the historical impact of CTT on the future of the global culture. Finally, I believe that computers and technology will end up having an indelible impact on the development of the human brain and the path of human evolution. Studies related to brain structure and human cognition are only just beginning to track changes related to individual use of technology, but it will take many decades before clear results might be shown.

Universal-to-Unique Continuum

The four samples in this study can be placed along Feldman's (1994) universal-to-unique continuum in relation to the development of the domain of computer technology. Individuals with CTT can progress through each of these stages as they develop their talent: (1) Universal, (2) cultural, (3) discipline-based, (4) idiosyncratic, and (5) unique (p. 23). Not only do these samples offer different perspectives, they also show different stages in the unidirectional development of individuals with CTT. Not every person in each group fit neatly into a category, but as a whole each sample was representative of a different stage on the continuum.

The participants in Sample 1 were the trendsetters for the computer science field, and many made unique contributions that went on to change the entire culture. Feldman (1994) pointed out how rare this occurrence is: "Only a tiny amount of novel thoughts are ever perceived as useful at all, even a tinier number continue to be useful for very long, and even a smaller number yet become continuing parts of an evolving field" (p. 47). To

185

have such a high concentration of these creative individuals in one place was unusual. All of the participants could be considered representative of the idiosyncratic stage of the continuum. They had learned the extent of the existing knowledge about programming and software design and then went beyond that to create something new. Some of the contributions did not last very long, while others made an enduring impact on society.

The computing adults interviewed in Sample 2 represented the discipline-based stage. The majority of them had jobs in the computer science field and had gained their knowledge and skills either with a college degree or through related work experience. They had made the choice to specialize their computer knowledge beyond the cultural stage. Some were content to stay at this level, while others chose to leave the field instead of moving forward. I believe that some of these participants have the potential to develop their talents to the idiosyncratic level within the domain of computer science. They have not yet reached their creative peak in their careers, and under the right circumstances they could progress. Based on Lehman's (1953) work on the most closely related field of mathematics, the majority of great mathematicians did not make their greatest contributions until their thirties and forties (p. 184). I do not know if it is possible for them to make a unique contribution comparable to Bill Gates, but the field continues to develop in tiny increments.

The longitudinal sample presented two different interview times which helped illustrate the development of the participants along the universal-to-unique continuum. Sample 3 was originally interviewed in high school where they were in the process of absorbing all the information that the technology culture had to offer. They also had their own peer culture within the computer programming club where they shared information

186

between themselves. They were eager to continue their quest for knowledge and taught themselves what they could about the discipline. The follow-up interviews were essential to show who made that full transition to the next level. At the second interview point, it was revealed that some participants continued into the discipline-based stage while others chose a different path. The transition from cultural to discipline-based required hard work, and those individuals who completed computer science degrees and got tech jobs have the potential to continue on to the highest level of the continuum.

Sample 4 required the most speculation when it came to their potential on the universal-to-unique continuum. Based on their age and experience, I would place them firmly in the cultural stage. These CTT students had a knack for absorbing information quickly and they took in everything around them related to computer technology. Researchers have examined how this next generation seems to be wired for technology use and have been labeled as *digital natives* (Prensky, 2001a; Prensky, 2001b). What makes these participants different from their age peers is their potential to continue along the continuum far above the average individual. Some of these teenagers taught themselves high level skills but they cannot be considered as part of the next stage until they gain experience at college or in a real world work setting. At that point some may choose to continue on or they may find they prefer a different line of work. Several of those gifted individuals have a high likelihood of demonstrating idiosyncratic thinking, but the researcher will have to wait for many years to see if that potential develops into CTT success.

Placed next to each other, these four samples are snapshots of the different stages of CTT development. Not only are these samples representative of different age groups

187

but they are also from different periods of time. Their interests and level of knowledge are reflected in this delineation. One important factor that affected the differences in abilities between groups was the *Flynn effect* (Rodgers, 1998). For over three generations IQ scores have been steadily rising, which would seem to indicate that each generation is getting smarter. What actually is happening is that the knowledge and skills that were considered specialized a decade or two before have trickled down into the general culture. This effect has happened particularly fast in relation to computers and technology. This phenomenon can be observed even at the highest levels of the continuum. Bill Gates from Sample 1 said, "People used to get Ph.D. theses for doing work that we now expect programmers to do as part of their jobs" (Lammers, 1989, p. 81). Therefore the types of skills and knowledge displayed by CTT individuals are expected to evolve, but their advancement along the universal-to-unique continuum will always be the same.

Limitations

This study of computer technology talent went through different phases before evolving into the final project. Initially I wanted to do an in-depth study on a single sample of students at a high tech secondary school, but that design turned out to not be feasible. The project was reformulated to look at several different samples and compare their experiences in relation to CTT development. This change ended up producing far richer data to support the research objectives of this study.

One limitation specific to Sample 3 was that I was not present at the first data collection point (Time 1). Unfortunately, the original audio recordings of the interviews were not available, and so the summarized transcripts and handwritten notes taken by the

188

original researcher had to be used. Overall, obtaining information from the other samples was relatively simple, but the task became harder with the youngest group, Sample 4. Parental permission was required for the students who were under 18-years-old, and convincing them to return the signature sheet before collecting the data proved to be difficult. Also because this study used interviews to collect responses instead of a simple survey, persuading students to spend an hour talking or chatting was challenging. There was only a 10% return rate on the permission slips, but the participants who responded proved to be helpful and informative.

This study followed an emergent design, and therefore did not have a specific theory structure for the data collection phase. Gagné's (2003) talent theory emerged as the best fit for the results after all the interview data had been obtained. If the researcher had started out with this developmental structure in mind, all of the interview questions could have been geared towards the different aspects present in that model. However, a completely structured study might have missed some of the more unique responses of the participants. Another way this research could have been different would have been to take a phenomenological approach. Instead of comparing multiple samples, a few select students could have been thoroughly studied in all aspects of their lives. If there had been limitless time, the researcher would have liked to shadow CTT students at school and spend time with their families at their homes to see how they used computers and technology on a daily basis.

This study was limited to the participants who volunteered for research, and it is possible that there were better representatives of CTT in the population who declined to be interviewed. The conclusions are based on the samples that the researcher had access to, and other samples could have yielded different results. An attempt was made to look at all aspects of talent development, and more benefit might be derived from focusing on a single aspect like creativity or motivation. Now that the full developmental path has been examined, each key characteristic can be studied more deeply. There was also no control sample with which to compare the responses of these participants, and therefore it was unknown how many of these same personal qualities could be found in the normal development of a person who chooses to use a computer. This research provided just one snapshot of the phenomenon of CTT.

One very important limitation was that the participants in this study were almost exclusively male, with the exception of one female in Sample 3. The general consensus in the existing literature appears to be that computer science is a male-dominated field, and the reasons behind that are still being speculated upon (McLester, 1998; Green, 2000; Macleod, D. Haywood, J. Haywood, & Anderson, 2002). Girls and women have dramatically increased their general computer usage during the last decade, but there is still a large gender gap within the discipline of computer science and engineering (CS/CE). In 2004, only 25% of the people who got a Master's degree in CS/CE were female, and only 17% of those who got a Bachelor's degree were female (*Computing Research Association*, 2005). At best that means that three-quarters of the population in the domain of computer science are male. Most of the participants did not comment on gender and computing, or if they did they simply said they did not know why there were so few girls in the field. More insight could have been gained into the effects of gender if an interview question about this topic had been asked of the participants. I will include this aspect of gender when developing my future research projects on this subject.

Implications

Knowing about the developmental path of students with computer technology talent can be beneficial to the field of gifted education as well as the field of computer technology. Hammersley (1998) emphasized that, "To be of value, research findings must not only be valid but also relevant to issues of actual or potential public concern" (p. 70). As technology continues to advance in society, the way in which CTT children are educated needs to reflect these changes. This research study is worthwhile to gifted students, teachers, parents, and school counselors whose primary job is to identify students for gifted services. Changes in both policy and practice are presented below, with suggestions for teachers and parents of CTT children.

Policy.

Part of the motivation behind the current research was spurred by the incorporation of two new categories (ICT and Design & Technology) into the United Kingdom's national curriculum for gifted and talented students (*Guidance on Teaching the Gifted and Talented,* 2003). Students in the UK are identified as gifted through testing, parent and pupil feedback, as well as teacher identification. In order to facilitate this process, the above website was launched in 2001 by the Qualifications and Curriculum Authority to provide example characteristics of gifted students in each subject area. Some sample identifiers in this area were, "Pupils demonstrate high levels of technological understanding and application," and, "Pupils transfer and apply ICT skills and techniques confidently in new contexts" (*Guidance on Teaching the Gifted and Talented,* 2003). These characteristics were useful as a frame of reference, and serve as an example of how policy changes could happen in the U.S. to the gifted curriculum.

The only current U.S. identification method for the phenomenon of computer technology talent (CTT) focuses exclusively on behavioral observation. A technology characteristics checklist was developed by Renzulli, et al. (2004) as one category of the Scales for Rating the Behavioral Characteristics of Superior Students (SRBCSS). In a later article, Siegle (2004) explained that the technology scale was, "Based on four key student characteristics: expertise using technology, interest and initiative in using technology, mentoring others in technology, and creative integration of technology" (p. 31). These behavioral characteristics were identified with generic sentences, such as, "The student demonstrates a wide range of technology skills," and, "The student assists other with technology related problems" (Renzulli, et al., 2004). I believe that students with CTT can be identified more effectively by expanding the assessment beyond a seven-question checklist. Without a practical computer interface for identifying students with high aptitude for technology, students who could benefit from gifted services could be overlooked by the educational system.

One possible means of CTT identification is an interactive computer literacy assessment developed by Educational Testing Services called *iSkills* (*ETS*, 2007). Although this test only covers one aspect of computer use (no programming or hardware knowledge is necessary), it requires the user to demonstrate their technology skills in a simulation of real world computer applications. I.R. Katz (2005) listed the seven student abilities related to ICT literacy that this test assessed: define, access, manage, integrate, evaluate, create, and communicate using computer technology (p. 4). This ETS test has been piloted in high schools and colleges to develop a standard technology literacy level, however it seems to be geared towards the minimum level of ICT knowledge necessary to graduate. There are no off-level norms for this assessment, and there is no differentiation for special populations yet. Since this is considered an upper level test, one way for educators to use this in a gifted program would be to administer this assessment to younger students who showed CTT potential. Continued development of authentic practical tests like these will benefit the study of students with computer technology talent.

The results of this current study showed the importance of looking at a child's talent in developmental terms during the identification process. Most of the participants argued against a one-time test to determine the level of someone's computer technology talent. Most identification practices focus almost exclusively on intellectual characteristics, but aspects like creative thinking and self-management can be just as important to a person's talent development. It is suggested to use interviews with students and parents that focus on their skills over time to help in the identification process. Johnsen (2004) wrote about how important interviews were to this initial stage:

> While many professionals advocate using interviews, little research can be found about how interviews fare in the identification of gifted students. In spite of the lack of research, interviews hold promise in the field for identifying students as gifted, especially those from low-income or culturally diverse backgrounds. (p. 27)

Students could also possibly show their counselor technology projects that they liked to work on outside of the school environment. Many of the participants disliked their school experiences, so they were less likely to display the full extent of their CTT skills in that environment. If only a behavioral scale is used for identification, then the pertinent behaviors might not be displayed at school. An identification interview that uses Gagné's

193

(2003) developmental model could uncover the bigger picture related to how these aspects synthesize to produce CTT in an individual.

The long-term goal of this research is to expand the state and national policy for gifted education to include provisions for students with computer technology talent. The participants in this study encouraged a change in policy. Lisa from the Sample 3 wrote, "I think a specific classification of gifted education [for CTT] could be helpful. Especially in identifying what school / programs should be allocated what funding or gear for those schools having a high number of technology talent students." Feldhusen (1996) advocated that, "The process of identifying talents should begin early and be viewed as a continuous process of ever better delineation of students' strengths, interests, and learning-growth styles" (p. 68). The goal is to successfully identify these children in order to qualify them for the district and state services that they deserve. Further research is needed to determine how the current educational policy could be modified to accommodate students with computer technology talent.

Practice.

Previous literature has focused solely on practical strategies for teachers to use in their classrooms with intellectually gifted students (Pyryt, Masharov, & Feng, 1993; Riley & Brown, 1997). Other articles emphasized the positive impact of using computers with gifted students and offered class activity suggestions for teachers (Belcastro, 2002; Morgan, 1993; Nugent, 2001; Steele, Battista, & Krockover, 1982). Although these program descriptions are beneficial to educators, there were few research studies on the effectiveness of these suggested lesson plans on the development of computer technology talent. One study by Sewell (1990) suggested that computers were an essential part of an enriched educational environment where students learned more effectively than in traditional settings. Based on this conclusion, he proposed that development could be accelerated and higher goals expected at an earlier age.

Educators are optimistic about the positive impact of technology in the general classroom, as Gardner (2000) predicted: "For the first time in human society, it is easy to envision a mass educational environment where instruction can be truly individual" (p. 13). This positive outlook was echoed be Dede (1998), who foresaw great strides being made at all levels of schooling:

> Successful technology-based innovations have the common characteristic that learners exceed everyone's expectations for what is possible. Second graders do fifth grade work; nine graders outscore twelfth grade students. What would those ninth graders be accomplishing if, from kindergarten on, they had continuous access to our best tools, curriculum, and pedagogy? (p. 11)

Computers can allow for personalized public education, and for those students whose talent is with technology, resources need to be provided at the appropriate level. Haskell from Sample 3 had a suggestion for a way the education system could be improved for all students: "Those with gifts for IT should be encouraged, while those with other gifts should be given opportunity to maintain a base 'technology literacy' included in a new 'liberal arts' curriculum." With the right balance of home and school conditions, these CTT children can grow up into successful adults, working to reach new technological heights in society.

At a presentation to the Kansas Association for the Gifted, Talented and Creative (KGTC), O'Brien and Friedman-Nimz (2006) elaborated on ways that teachers could encourage computer technology talent in their students. After an open discussion with a group of Kansas certified gifted teachers, several suggestions for CTT students were listed such as authentic computer projects, accelerated technology classes, a compacted computer curriculum, connecting with technology mentors in the community, and getting the students involved with working on the school's website. More research needs to be done that is specifically focused on the effectiveness of different teacher/student interactions in relation to CTT activities in the school. Facer, J. Furlong, R. Furlong, and Sutherland (2003) argued that, "Schools and teachers are key resources in expanding young people's access to the cultural, social, technological and knowledge tools beyond those available in their homes and immediate cultural contexts" (p. 203). As much as the younger generation already knows about computers, what they do with that technology should be applied to challenge and educate them in an interactive environment.

Research has shown that parents and other adults play a key role in encouraging a child's computer use to go deeper than video games and internet surfing (Attewell & Battle, 1999). As part of a nationwide study in the UK, Livingstone (2001) interviewed and observed 30 families and found that the level of parental involvement largely depended on the location of the computer in the home and whether it had to be shared with other members of the family. If the machine was placed in a common area such as the kitchen or hallway, the likelihood that parents would get involved with their child's computer activities increased. Parents should encourage their children to pursue their interests in computers as well as establish household rules that reflect their own values and beliefs. Rimm (1994) wrote that parents should provide emotional support and praise, but should allow their children to initiate activities based on their internal motivation instead of for external rewards. Porter (2005) stated, "My belief about our role as parents is not to impose our own ambitions on our children, but to help them become who they

196

need to be in life" (p. 208). Parents with gifted children need to trust their little self-directed learners and allow them to explore computers and technology in a supportive home environment.

For extraordinarily gifted children, parents should do what they can to acquire the resources necessary for them to expand their children's knowledge and skills. Perino and Perino (1981) talked about the importance of parents in making decisions affecting their child's education, such as acceleration. The participants in this study who were able to attend college courses while still in high school said they had a positive experience, and so parents should inquire about similar services in their own districts. In Lehman's (1953) book *Age and Achievement*, he determined that the earlier extraordinarily creative people started in their field, the more and better creative work they produced. It has yet to be seen the wider impact of precocious computer prodigies who started at extremely young ages, and this is a line of study that should be explored further. Until more is known, parents should continue to be encouraging and supportive of their children in all talent development areas.

Future Research

In the seminal book, *Developing Talent in Young People*, Bloom (1985) specifically explored talent development in sports, music, art, and mathematics. In his conclusion he conceded that research needed to be conducted in other areas of talent to find out more about the development process. The domain of computer technology has advanced enough that another in-depth study similar to Bloom's could be conducted into the lives and achievements of CTT children.

After mapping the developmental path of gifted students with computer

197

technology talent, the next stage of research could compare this to the normal development of an average child. The focus of Feldman's (1994) research was on nonuniversal development, but a study has yet to be conducted with a side-by-side comparison of the development of computer skills in a typical student with a CTT student. The same research guide from this study could be followed to produce results about the standard development of an average person growing up in the digital era. Another avenue of research that needs exploring is a comparison of students from different populations (e.g. race, SES, natural-born vs. immigrants, etc.). Research studies often focus on the easily accessible white middle-class suburban student, this one included, and research can always be expanded in this area.

There was only one female participant in this study, and the effect of gender on CTT is another perspective that merits further exploration. The small amount of research on gifted girls and technology suggested that if they are discouraged from using computers at an early age, then they are at risk of missing out on developing some of the advanced problem-solving skills promoted by computer use (Berger, 2003). Several strong female voices in the computer world encourage teachers to actively involve girls in more computing activities in elementary school, which can help them incorporate technology skills into their sense of self at an early age (Ettenheim, Furger, Siegman, & McLester, 2000). The specific development of gifted girls in relation to computer technology is an area where I would like to see more research conducted.

The research area of talent development continues to grow, and Van Tassel-Baska (2001) pointed out that researchers still do not know the answer to questions such as, "What intervention works best with what students at a given age/stage of development?"

(p. 25). Coleman (2001) agreed with Van Tassel-Baska and added another question about

what types of environments are best for these students:

> I had come to believe that research approaches that focus on the individual out of the context in which giftedness grows, or that look at the person before looking at the setting in which giftedness grows, were conceived in a backward fashion. (p. 164)

Therefore different interventions for CTT students should be studied at different points in

their lives to evaluate the appropriateness and effectiveness. Teaching strategies, school

policies, and mentoring opportunities are practices that should be researched in relation to

the development of computer technology talent.

Summary

Computer technology talent has just recently emerged as an area of interest in

gifted education (Siegle, 2005). The developmental path a person takes towards CTT is

affected by their natural abilities, intrapersonal catalysts, and environmental influences.

According to Gagné (2003), only the top 10% of gifted individuals who start out with a

natural gift in this area will be able to successfully develop their skills into talents. The

four samples in this study showed a pattern of distinct characteristics related specifically

to computers and technology. More research is needed to determine the distinctions

between CTT and other related talent areas. Alternative assessments and resources need

to be made available in schools to recognize this new way of thinking in the digital age.

Every student deserves to learn in the best environment possible, and accommodations

for students with CTT should not be overlooked. The future contributions to society that

these talented individuals could make in the field of computing are unknown, therefore

they should be supported in their technology endeavors both in school and at home.

References

Abbate, J. (1999). *Inventing the internet.* Cambridge, MA: The MIT Press.

Armstrong, A., & Casement, C. (2000). *The child and the machine: How computers put our children's education at risk.* Beltsville, MD: Robins Lane Press.

Arskey, H., & Knight, P. (1999). *Interviewing for social scientists.* London: Sage Publications, Inc.

Assouline, S.G. (2003). Psychological and educational assessment of gifted children. In N. Colangelo & G.A. Davis (Eds.) *Handbook of Gifted Education*, (3rd ed., pp. 124-145). Boston: Allyn & Bacon.

Attewell, P., & Battle J. (1999). Home computers and school performance. *The Information Society, 15(1),* 1-10.

Attewell, P., Suazo-Garcia, B., & Battle, J. (2003). Computers and young children: Social benefit or social problem? *Social Forces, 82(1),* 277-296.

Bamberger, J. (1982). Growing up prodigies: The midlife crisis. In D.H. Feldman (Ed.), *Developmental approaches to giftedness and creativity,* (pp. 61-77). San Francisco: Jossey-Bass.

Belcastro, F. (2002). Electronic technology and its use with rural gifted students. *Roeper Review, 25(1),* 14-16.

Berger, S. (2003). Surfing the Net: Are we dumbing down our daughters? *Understanding Our Gifted, 15(2),* 1040-1350.

Bloom, B. S. (1985). (Ed.) *Developing talent in young people.* New York: Simon & Schuster.

Bowen, S., Shore, B.M., & Cartwright, G.F. (1992). Do gifted children use computers differently? A view from 'the factory'. *Gifted Education International, 8(3),* 151-154.

Bozionelos, N. (2001). Computer interest: A case for expressive traits. *Personality and Individual Differences, 33(3),* 427-444.

Bransford, J.D., & Stein, B.S. (1984). *The ideal problem solver: A guide for improving thinking, learning, and creativity.* New York: W.H. Freeman & Co.

Brantley, M.E., & Coleman, C. (2001). *Winning the technology talent war: A manager's guide to recruiting and retaining tech workers.* New York: McGraw-Hill.

Braun, L. (1999). From misfits to software pioneers. *Technology & Learning, 20(5),* 58.

Bryce, J. (2001). The technological transformation of leisure. *Social Science Computer Review, 19(1),* 7-16.

Bulls, M. R., & Riley, T.L. (1997). Weaving qualitatively differentiated units with the World Wide Web. *Gifted Child Today, 20(1),* 20-27.

Campbell, R.L. (1990). Can there be developmental psychology of human-computer interaction? *Contemporary psychology, 35 (4),* 381-383.

Campbell, R. L., Brown, N. R., & DiBello, L. A. (1992). The programmer's burden: Developing expertise in computer programming. In R. R. Hoffman (Ed.), *The Psychology of Expertise: Cognitive Research and Empirical AI,* (pp. 269-294). New York: Springer.

Chandler, S. (2005, October). *A report from the digital contact zone: Differences between insider and newcomer mindsets in composition and literacy studies.* Paper presented at the Association of Internet Researchers, Chicago, IL.

Chiero, R., Sherry, L., Bohlin, R., & Harris, S. (2003). Increasing comfort, confidence, and competence in technology infusion with learning communities. *TechTrends, 47(2),* 34-38.

Clark, A. (2003). *Natural born cyborgs: Minds, technologies, and the future of human intelligence.* New York: Oxford University Press.

Cohen, V.L. (2001). Learning styles and technology in a ninth-grade high school population. *Journal of Research on Technology in Education, 33(4),* 355-366.

Coleman, L.J. (2001). A 'rag quilt': Social relationships among students in a special high school. *Gifted Child Quarterly, 45(3),* 164-173.

Coleman, L.J., & Cross, T.L. (1988). Is being gifted a social handicap? *Journal for the Education of the Gifted, 11(4),* 41-56.

Coleman, L.J., Guo, A., & Dabbs, C.S. (2007). The state of qualitative research in gifted education as published in American journals. *Gifted Child Quarterly, 51(1),* 51-63.

Coleman, L.J., Sanders, M.D., & Cross, T.L. (1997). Perennial debates and tacit assumption in the education of gifted children. *Gifted Child Quarterly, 41(3),* 105-111.

Collis, B., & Anderson, R. (1994). Computer literacy for the 1990s: Theoretical issues for an international assessment. *Computers in the Schools, 11(2),* 55-72.

Computer History Museum. (2006). Retrieved on June 27, 2007, from http://www.computerhistory.org/timeline/?category=cmptr

Computing Research Association (2005). Taulbee trends: Female students & faculty. Retrieved July 21, 2007, from http://www.cra.org/info/taulbee/women.html

Costa, A.L., & Kallick, B. (2000). *Assessing and reporting habits of mind*. Alexandra, VA: Association for Supervision and Curriculum Development.

Cross, T.L. (2005). Nerds and geeks: Society's evolving stereotypes of our students with gifts and talents. *Gifted Child Today, 28(4)*, 26-27.

Cross, T.L. (1994). Alternative inquiry and its potential contributions to gifted education: A commentary. *Roeper Review, 16(4)*, 284-285.

Csikszentmihalyi, M. (1996). *Creativity: Flow and the psychology of discovery and invention*. New York: HarperCollins Publishers, Inc.

Csikszentmihalyi, M., Rathunde, K., & Whalen, S. (1993). *Talented teenagers: The roots of success and failure*. Cambridge: Cambridge University Press.

Csikszentmihalyi, M. & Robinson, R.E. (1986). Culture, time, and the development of talent. In R.J. Sternberg & J.E. Davidson (Eds.), *Conceptions of Giftedness*, (pp. 264-284). Cambridge: Cambridge University Press.

Cuban, L. (2001). *Oversold and underused: Computers in the classroom*. Cambridge: Harvard University Press.

Custer, R.L. (1995). Examining the dimensions of technology. *International Journal of Technology and Design Education, 5(3)*, 219-244.

Dabrowski, K., & Piechowski, M.M. (1977). *Theory of levels of emotional development*, Vols. 1 & 2. Oceanside, NY: Dabor Science.

Davis, G.A. (2003). Identifying creative students, teaching for creative growth. In N. Colangelo & G.A. Davis (Eds.) *Handbook of Gifted Education* (3rd ed., pp. 311-324). Boston: Allyn & Bacon.

Dede, C. (in press). *Six challenges for educational technology*. Retrieved June 28, 2007, from http://www.virtual.gmu.edu/SS_research/cdpapers/ascdpdf.htm

Dreyfus, H., & Dreyfus, S. (1986). *Mind over machine*. New York: The Free Press.

Durden, W.G., & Tangherlini, A.E. (1993). *Smart kids: How academic talents are developed and nurtured in America.* Toronto: Hogrefe & Huber Publishers.

Eisner, E.W., & Peshkin, A. (1990). (Eds.) *Qualitative inquiry in education: The continuing debate*. New York: Teachers College Press.

Elementary and Secondary Education Act. (2001). Retrieved May 29, 2007, from http://www.ed.gov/policy/elsec/leg/esea02/index.html

Ericsson, K.A., & Smith, J. (1991). *Toward a general theory of expertise: Prospects and limits*. New York: Cambridge University Press.

ETS (2007). iSkills - Information and communication technology literacy test. Retrieved July 21, 2007, from http://www.ets.org/portal/site/ets/ menuitem.435c0b5cc7bd0ae7015d9510c3921509

Ettenheim, S., Furger, R., Siegman, L., & McLester, S. (2000). Tips for getting girls involved. *Technology & Learning, 20(8),* 34-36.

Facer, K., Furlong, J., Furlong, R., & Sutherland, R. (2003). *Screenplay: Children and computing in the home*. New York: RoutledgeFalmer.

Feldhusen, J.F. (2005). Giftedness, talent, expertise, and creative achievement. In R.J. Sternberg & J.E. Davidson (Eds.), *Conceptions of Giftedness* (2nd ed., pp. 64-79). Cambridge: Cambridge University Press.

Feldhusen, J.F. (1996). How to identify and develop special talents. *Educational Leadership, 53(5),* 66-69.

Feldhusen, J.F. (1995). Talent development: The new direction in gifted education. *Roeper Review, 18(2)*, 92.

Feldman, D.H. (2000). Developmental theory and the expression of gifts and talents. In C. Van Lieshout & P. Heymans (Eds). *Developing Talent Across the Life Span*, (pp. 3-16). East Sussex, UK: Psychology Press.

Feldman, D.H. (1994). *Beyond universals in cognitive development* (2nd ed.). Norwood, NJ: Ablex Publishing Corporation.

Feldman, D.H., & Benjamin, A. C. (1986). Giftedness as a developmentalist sees it. In R.J. Sternberg & J.E. Davidson (Eds.), *Conceptions of Giftedness*, (pp. 285-305). Cambridge: Cambridge University Press.

Feldman, D.H., Csikszentmihalyi, M., & Gardner, H. (1994). *Changing the world: A framework for the study of creativity*. Westport, CT: Greenwood Press.

Feldman, D.H., & Goldsmith, L.T. (1986). *Nature's gambit: Child prodigies and the development of human potential*. New York: Basic Books, Inc.

Flores, J.G., & Alonso, C.G. (1995). Using focus groups in educational research. *Evaluation Review, 19(1)*, 84-101.

Fontana, A., & Frey, J.H. (2003). The interview: From structured questions to negotiating text. In N. Denzin & T. Lincoln (Eds.), *Collecting and Interpreting Qualitative Materials*, (2nd ed., pp. 61-106) Thousand Oaks, CA: Sage Publications, Inc.

Fox, S., & Madden, M. (2006). Generations online. *Pew Internet & American Life Project*. Retrieved May 5, 2006 from http://www.pewinternet.org/pdfs/PIP_Generations_Memo.pdf

Frauenheim, E. (2005). Can Johnny still program? *CNET News.com*, Retrieved June

20, 2007 from http://news.com.com/Can+Johnny+still+program/2008-1036_3-5675770.html

Freeman, J. (2000). Teaching for talent: Lessons for the research. In C. Van Lieshout & P. Heymans (Eds). *Developing Talent Across the Life Span*, (pp. 231-248). East Sussex, UK: Psychology Press.

Friedman-Nimz, R., Lacey, J., & Denson, D. (2002). Teaming with G/T students: Improving online teacher education. *The KGTC Bugle* (4), 4-5.

Friedman-Nimz, R., & Skyba, O. (in press). Personality qualities that help or hinder gifted and talented individuals. In L.V. Shavinina (Ed.), *International Handbook on Giftedness*.

Gagné, F. (2003). Transforming gifts into talents: The DMGT as a developmental theory. In N. Colangelo & G.A. Davis (Eds.) *Handbook of Gifted Education*, (3rd ed., pp. 60-74). Boston: Allyn & Bacon.

Gagné, F. (1999). *Tracking talents: Identifying multiple talents through peer, teacher, and self-nomination.* Waco, TX: Prufrock Press, Inc.

Gallagher, J. J. (2003). Issues and challenges in the education of gifted students. In N. Colangelo & G.A. Davis (Eds.) *Handbook of Gifted Education* (3rd ed., pp. 11-23). Boston: Allyn & Bacon.

Gardner, H. (2000). The complete tutor. *Technos, 9(3),* 10-13.

Gardner, H. (1999). *Intelligence reframed: Multiple intelligences for the 21st century.* New York: Basic Books.

Gardner, H. (1983). *Frames of mind: The theory of multiple intelligences.* New York: Basic Books.

Gee, J.P. (2003). *What video games have to teach us about learning and literacy.* New

    York: Paulgrave MacMillan.

Giacquinta, J.B., Bauer, J.A., & Levin, J.E. (1993). *Beyond technology's promise:*

    An examination of children's educational computing at home. Cambridge:

    Cambridge University Press.

Gill, T. (1996). *Electronic children: How children are responding to the information*

    revolution. London: National Children's Bureau Enterprises.

Gillespie, C.W., & Beisser, S. (2001, August). Developmentally appropriate LOGO

    computer programming with young children. *Information Technology in*

    Childhood Education, 1, 229-245.

Glaser, R. (1996). Changing the agency for learning: Acquiring expert performance.

    In K.A. Ericsson (Ed.) *The Road to Excellence: The Acquisition of Expert*

    Performance in the Arts and Sciences, Sports and Games, (pp. 303-311).

    Mahwah, NJ: Lawrence Erlbaum Associates, Inc.

Goldsmith, L.T. (1990). The Timing of talent: The facilitation of early prodigious

    achievement. In M. Howe (Ed.) *Encouraging the Development of Exceptional*

    Skills and Talents, (pp. 17-31). Leicester, UK: British Psychological Society.

Gottfredson, L.S. (2003). The science and politics of intelligence in gifted education.

    In N. Colangelo & G.A. Davis (Eds.) *Handbook of Gifted Education,* (3rd

    ed., pp. 24-40). Boston: Allyn & Bacon.

Gow, P. (2004). Technology & the culture of learning: How our digital tools change

    the nature of school. *Independent School, 63(4),* 18-26.

Gowan, J.C., Khatena, J., & Torrance, E.P. (1979). *Educating the ablest: A book of*

readings on the education of gifted children. Itasca, IL: F.E. Peacock
Publishers, Inc.

Graham, P. (2004). *Hackers and painters: Big ideas from the computer age.*
Sebastopol, CA: O'Reilly Media, Inc.

Graham, P. (2003). Why nerds are unpopular. Retrieved June 28, 2007, from
http://paulgraham.com/nerds.html

Green, M. (2000). Why aren't girls more tech savvy? *NEA Today, 19(3),* 31.

Greenfield, P.M. (1984). *Mind and media: The effects of television, video games, and
computers.* Cambridge, MA: Harvard University Press.

Guidance on Teaching the Gifted and Talented. (2003). Retrieved April 28, 2004,
from http://www.nc.uk.net/gt/index.htm

Hammersley, M. (1998). Standards for assessing ethnographic research. *Reading
Ethnographic Research* (2nd ed.). London: Longman.

Hardin, D. E. (1989). From gifted virgins to talented nerds. *Gifted Child Today,
12(6),* 45-47.

Healy, J. (1998). *Failure to connect: How computers affect our children's minds - for
better and worse.* New York: Simon and Schuster.

Heaney, L. (2003). Facing the challenges: Using information and communications
technology to support teaching and learning. *Gifted Education International,
17(1),* 59-72.

Heinzen, T.E. (in press). What do we know about computer technology talent? In
L.V. Shavinina (Ed.), *International Handbook on Giftedness.*

Hellenga, K. (2002). Social space, the final frontier: Adolescents on the internet. In

   J.T. Mortimer & R.W. Larson (Eds.) *The Changing Adolescent Experience:*

   Societal Trends and the Transition to Adulthood, (pp. 208-249). Cambridge:

   Cambridge University Press.

Hennessey, B.A., & Amabile, T.M. (1988). The conditions of creativity. In R.J.

   Sternberg (Ed.) *The Nature of Creativity: Contemporary Psychological*

   Perspectives, (pp. 11-38). Cambridge, MA: Cambridge University Press.

Hertzfeld, A. (1982). Gobble, gobble, gobble. *Folklore.org*, Retrieved on June 28,

   2007, from http://www.folklore.org/StoryView.py?project=Macintosh&story=

   Gobble_Gobble_Gobble.txt

Holland, J.L. & Astin, A.W. (1962). The prediction of the academic, artistic, scientific,

   and social achievement of undergraduates of superior scholastic aptitude. *Journal*

   of Educational Psychology, 53, 132-143.

Holloway, S.L., & Valentine, G. (2003). *Cyberkids: Children in the information age.*

   New York: RoutledgeFalmer.

Horowitz, F.D. (1987). A developmental view of giftedness. *Gifted Child Quarterly,*

   31(4), 165-168.

Horrigan, J.B. (2007). A typology of information and communication technology

   users. *Pew Internet & American Life Project.* Retrieved May 8, 2007 from

   http://www.pewinternet.org/pdfs/PIP_ICT_Typology.pdf

Horrigan, J.B. (2000). New internet users: What the do online, what they don't, and

   implications for the 'net's future. *Pew Internet & American Life Project.*

   Retrieved June 30, 2007, from http://www.pewinternet.org/pdfs/

New_User_Report.pdf

Howe, N., & Strauss, W. (2000). *Millennials rising: The next great generation.* New
York: Vintage Books.

International Society for Technology in Education (ISTE) (2005). National
educational technology standards for students. Retrieved July 22, 2007, from
http://cnets.iste.org/students

Jenkins-Friedman, R., & Tollefson, N. (1991). Resiliency in cognition and motivation: Its
applicability to giftedness. In N. Colangelo, S.G. Assouline, & D.L. Ambroson
(Eds.). *Talent Development: Proceedings from the 1991 Henry B. and Jocelyn
Wallace National Research Symposium on Talent Development,* (pp. 325-333).
Unionville, NY: Trillium Press.

Johnsen, S. K. (2004). *Identifying gifted students: A practical guide.* Waco, TX:
Prufrock Press, Inc.

Johnson, S. (2005). *Everything bad is good for you: How today's popular culture is
actually making us smarter.* New York: Riverhead.

Johnson, S. (1997). *Interface culture: How new technology transforms the way we
create and communicate.* New York: HarperCollins.

Jordan, A.B. (2002). A family systems approach to examining the role of the internet
in the home. In S.L. Calvert, A.B. Jordan, & R.R. Cocking (Eds.). *Children
in the Digital Age: Influences of Electronic Media on Development,* (pp. 231-
247). Westport, CT: Praeger.

Katz, I.R. (2005). Beyond technical competence: Literacy in information and communication technology. *Educational Testing Service (ETS)*. Retrieved July 17, 2007, from http://www.ets.org/Media/Tests/ICT_Literacy/pdf/ ICT_Beyond_Technical_Competence.pdf

Katz, J. (2000). *Geeks: How two lost boys rode the internet out of Idaho*. New York: Villard Books.

Kerr, B., & Cohn, S.J. (2001). *Smart boys: Talent, manhood, and the search for meaning*. Scottsdale, AZ: Great Potential Press, Inc.

Kidder, T. (1981). *The soul of a new machine*. Boston: Little, Brown and Co.

Kinney, D. A. (1993). From nerds to normals: The recovery of identity among adolescents from middle school to high school. *Sociology of Education, 66(1),* 21-40.

Kansas State Department of Education (2004). Student support services – Gifted education services. Retrieved July 17, 2007, from http://www.kansped.org/ksde/gifted/giftedindex.html

Kurzweil, R. (2001, April). As machines become more like people, will people become more like God? *Talk*, 152-155.

Lammers, S. (1989). *Programmers at work: Interviews with 19 programmers who shaped the computer industry*. Redmond, WA: Tempus Books.

Landrum, G.N. (1993). *Profiles of genius: Thirteen creative men who changed the world*. Buffalo, NY: Prometheus Books.

Lawler, R.W. (1985). *Computer experience and cognitive development: A child's learning in a computer culture*. New York: John Wiley & Sons.

Lee, C-M., Miller, W.F., Hancock, M.G., & Rowen, H.S. (2000) (Eds). *The Silicon*

Valley edge: A habitat for innovation and entrepreneurship. Palo Alto, CA: Stanford University Press.

Lehman, H.C. (1953). *Age and achievement*. Princeton, NJ: Princeton University Press.

Lenhart, A., Lewis, O., & Rainie, L. (2001). Teenage life online: The rise of the instant-message generation and the internet's impact on friendships and family relationships. *Pew Internet & American Life Project*. Retreived June 28, 2007, from http://www.pewinternet.org/pdfs/PIP_Teens_Report.pdf

Leu, D. J., Zawilinski, L., Castek, J., Banerjee, M., Housand, B., Liu, Y., & O'Neil, M. (in press). What is new about the new literacies of online reading comprehension? In A. Berger, L. Rush, & J. Eakle (Eds.). *Secondary school reading and writing: What research reveals for classroom practices*. National Council of Teachers of English/National Conference of Research on Language and Literacy: Chicago, IL.

Lindlof, T.R., & Taylor, B.C. (2002). *Qualitative communication research methods* (2nd ed.). Thousand Oaks, CA: Sage Publications, Inc.

Livingstone, S. (2001). Children on-line: Emerging uses of the internet at home. *Journal of the IBTE, 2(1),* 1-7.

Machlowitz, M. (1985). *Whiz kids: Success at an early age*. New York: Arbor House.

Macleod, H., Haywood, D., Haywood, J., & Anderson, C. (2002). Gender & information and communications technology: A 10-year study of new undergraduates. *TechTrends, 46(6),* 11-15.

MacPherson, R.T. (1998). Factors affecting technological trouble shooting skills. *Journal of Industrial Teacher Education, 35(4),* 5-28.

Maslow, A. (1998). *Toward a psychology of being*, (3rd ed.). New York: Wiley.

Maslow, A. (1991). Critique of self-actualization theory. *Journal of Humanistic Education and Development, 29(3)*, 103-108.

McHugh, J. (2005). Synching up with the iKid: Educators must work to understand and motivate a new kind of digital learner. *Edutopia, 1(7)*, 33-35.

McLester, S. (1998). Girls and technology: What's the story. *Technology and Learning, 19(3)*, 18-26.

Meszaros, P.S. (2004). The wired family: Living digitally in the postinformation age. *American Behavioral Scientist, 48(4)*, 377-390.

Morelock, M. J., & Feldman, D.H. (2003). Extreme precocity: Prodigies, savants, and children of extraordinarily high IQ. In N. Colangelo & G.A. Davis (Eds.) *Handbook of Gifted Education*, (3rd ed., pp. 455-469). Boston: Allyn & Bacon.

Morgan, T.D. (1993). Technology: An essential tool for gifted & talented education. *Journal for the Education of the Gifted, 16(4)*, 358-371.

N6 (NUD*IST) Qualitative Data Analysis Software (Version 6.0) (2002). [Computer software]. Doncaster, Australia: QSR International Pty Ltd.

Nugent, S.R. (2001). Technology and the gifted: Focus, facets, and the future. *Gifted Child Today, 24(4)*, 38-45.

O'Brien, B., & Friedman-Nimz, R. (2006, October). *You and your computer geek: Advocating and nurturing*. Paper presented at the annual meeting of the Kansas Association for the Gifted, Talented and Creative, Lawrence, KS.

O'Brien, B., Friedman-Nimz, R., Lacey, J., & Denson, D. (2005). From bits and bytes
to C++ and websites: What is computer talent made of? *Gifted Child Today,
28(3),* 56-64.

Oppenheimer, T. (2003). *The flickering mind: The false promise of technology in the
classroom and how learning can be saved.* New York: Random House.

Organization Statistics. (2007). Retrieved May 31, 2007 from
http://www3.ksde.org/cgi-bin/dist_rpt_yrs?org_no=D0497

Orleans, M., & Laney, M. (2000). Children's computer use in the home: Isolation or
sociation? *Social Science Computer Review, 18(1),* 56-72.

O'Tuel, F.S. (1991). Developmental and longitudinal research. In N.K. Buchanan &
J.F. Feldhusen (Eds.) *Conducting Research and Evaluation in Gifted
Education: A Handbook of Methods and Applications*, (pp. 95-113). New
York: Teachers College Press.

Papert, S. (1993). *The children's machine: Rethinking school in the age of the
computer.* New York: Basic Books.

Papert, S. (1980). *Mindstorms: Children, computers, and powerful ideas.* New York:
Basic Books.

Perino, S.C., & Perino, J. (1981). *Parenting the gifted: Developing the promise.* New
York: R.R. Bowker Company.

Pew Internet & American Life Project (2007). Internet adoption. Retrieved June 30,
2007, from http://www.pewinternet.org/trends/Internet_Adoption_4.26.07.pdf

Piaget, J. (1972). Intellectual evolution from adolescence to adulthood. *Human
Development, 15,* 1-12.

Piaget, J. (1952). *The origins of intelligence in children.* New York: International

Universities Press.

Piirto, J. (1999). *Talented children and adults: Their development and education.*

(2nd ed.). Upper Saddle River, NJ: Prentice-Hall, Inc.

Plowman, L., & Stephen, C. (2003). A 'benign addition'? Research on ICT and pre-

school children. *Journal of Computer Assisted Learning, 19,* 149-164.

Porter, L. (2005). *Gifted young people: A guide for teachers and parents* (2nd ed.).

New York: Open University Press.

Prensky, M. (2001a). Digital natives, digital immigrants. *On the Horizon, 9(5),* NCB

University Press.

Prensky, M. (2001b). Digital natives, digital immigrants, part II: Do the really think

differently? *On the Horizon, 9(6),* NCB University Press.

Pyryt, M.C. (2003). Technology and the gifted. In N. Colangelo & G.A. Davis (Eds.)

Handbook of Gifted Education, (3rd ed., pp. 582-589). Boston: Allyn&Bacon.

Pyryt, M.C., Masharov, Y.P., & Feng, C. (1993). Programs and strategies for

nurturing talents/gifts in science and technology. In K.A. Heller, F.J. Monks,

& A.H. Passow (Eds.*), International Handbook of Research and Development

of Giftedness and Talent* (pp. 453-471). Oxford: Pergamon.

Rainie, L., & Packel, D. (2001). More online, doing more. *Pew Internet & American

Life Project*. Retrieved June 30, 2007, from http://www.pewinternet.org/pdfs/

PIP_Changing_Population.pdf

Rathunde, K. (1992). *Playful and serious interest: Two faces of talent development in adolescence.* Paper presented at the 1991 Henry B. Jocelyn Wallace National

Research Symposium on Talent Development, Unionville, NY.

Renzulli, J.S. (2003). Conception of giftedness and its relationship to the development of

social capital. In N. Colangelo & G.A. Davis (Eds.) *Handbook of Gifted*

Education, (3rd ed., pp. 75-87). Boston: Allyn & Bacon.

Renzulli, J.S., & Purcell, J.H. (1996). Gifted education: A look around and a look

ahead. *Roeper Review, 18(3)*, 173-178.

Renzulli, J.S., Smith, L. H., White, A.J., Callahan, C.M., Hartman, R.K., Westberg,

K.L., Gavin, K.M., Reis, S.M., Siegle, D., & Sytsma, R.E. (2004). *Scale for*

rating the behavioral characteristics of superior students. Mansfield Center,

CT: Creative Learning Press, Inc.

Rideout, V., Roberts, D. F., & Foehr, U. G. (2005). *Generation M: Media in the lives*

of 8-18 Year-Olds. Washington, DC: Kaiser Family Foundation.

Riley, T.L., & Brown, M.E. (1997). Computing for clever kids: The future is what we

make it. *Gifted Child Today, 20(5),* 22-29.

Rimm, S.B. (1994). *Keys to parenting the gifted child.* New York: Barron's Educational

Series, Inc.

Rodgers, J.L. (1998). A critique of the Flynn Effect: Massive IQ gains, methodological

artifacts, or both? *Intelligence, 26(4)*, 337-356.

Rossman, G.B., & Rallis, S. F. (2003). *Learning in the field: An introduction to*

qualitative research. Thousand Oaks, CA: Sage Publications, Inc.

Rushkoff, D. (1999). *Playing the future: What we can learn from digital kids.* New

York: Riverhead Trade.

Salomon, G., Perkins, D., & Globerson, T. (1991). Partners in cognition: Extending

   human intelligence with intelligent technologies. *Educational Researcher, 20(3),*

   2-9.

Saunders, D., & Thagard, P. (2005). Creativity in computer science. In J.C. Kaufman

   & J. Baer (Eds.) *Creativity Across Domains: Faces of the Muse,* (pp. 153-

   167). Mahwah, NJ: Lawrence Erlbaum Associates, Publishers.

Schneider, B.H. (1987). *The gifted child in peer group perspective.* New York:

   Springer-Verlag New York, Inc.

Schultz, R.A., & Delisle, J.R. (2003). Gifted adolescents. In N. Colangelo & G.A. Davis

   (Eds.) *Handbook of Gifted Education,* (3rd ed., pp. 483-492). Boston: Allyn &

   Bacon.

Schwartau, W. (2001). *Internet & computer ethics for kids (and parents & teachers*

   who haven't got a clue). Seminole, FL: Winn Schwartau & Interpact, Inc.

Scriptol.org. (2007). Alphabetical list of programming languages. Retrieved June 30,

   2007, from http://www.scriptol.org/alphabetical-programming-language-list.html

Selwyn, N. (2005). The social processes of learning to use computers. *Social Science*

   Computer Review, 23(1), 122-135.

Sewell, D. (1990). *New tools for new minds: A cognitive perspective on the use of*

   computers with young children. New York: St. Martin's Press.

Shavinina, L.V. (in press). On entrepreneurial giftedness. In L.V. Shavinina (Ed.),

   International Handbook on Giftedness.

Siegle, D. (2005). *Using media & technology with gifted learners.* Waco, TX:

   Prufrock Press.

Siegle, D. (2004). Identifying students with gifts and talents in technology. *Gifted Child Today, 27(4)*, 30-33, 64.

Silverman, D. (2005). *Doing qualitative research* (2nd ed.). Thousand Oaks, CA: Sage Publications, Inc.

Smith, J.K., & Deemer, D.K. (2003). The problem of criteria in the age of relativism. In N. Denzin & T. Lincoln (Eds.), *Collecting and Interpreting Qualitative Materials*, (2nd ed., pp. 427-457) Thousand Oaks, CA: Sage Publications, Inc.

Smith, G.G., & Grant, B. (2000). From players to programmers: A computer game design class for middle-school children. *Journal of Educational Technology Systems, 26(3)*, 263-275.

Steele, K., Battista, M., & Krockover, G. (1982). The effect of microcomputer assisted instruction upon the computer literacy of high ability students. *Gifted Child Quarterly, 26(4)*, 162-164.

Stoll, C. (2000). *High-Tech Heretic*. New York: Anchor Books.

Stonier, T., & Conlin, C. (1985). *The three Cs: Children, computers, and communication.* New York: John Wiley & Sons.

Strot, M. (1999). Recreational computing. *Gifted Child Today, 22(5)*, 36-37.

Subotnik, R.F. (1993). Generate and test: The creative world of Joseph Bates. *Journal for the Education of the Gifted, 16(3)*, 311-322.

Subotnik, R. F., Olszewski-Kubilius, P., & Arnold, K.D. (2003). Beyond Bloom: Revisiting environmental factors that enhance or impede talent development. In J.H. Borland (Ed.) *Rethinking Gifted Education*, (pp. 227-238). New York: Teachers College Press.

Subrahmanyam, K., Greenfield, P.M., Kraut, R., & Gross, E. (2002). The impact of

    computer use on children's and adolescents' development. In S.L. Calvert,

    A.B. Jordan, & R.R. Cocking (Eds.) *Children in the Digital Age: Influences*

    of Electronic Media on Development, (pp. 3-33). Westport, CT: Praeger.

Suran, B.G. (1978). *Oddballs: The social maverick and the dynamics of individuality.*

    Chicago: Nelson-Hall Inc. Publishers.

Tannenbaum, A.J. (2003). Nature and nurture of giftedness. In N. Colangelo & G.A.

    Davis (Eds.) *Handbook of Gifted Education,* (3rd ed., pp. 45-59). Boston: Allyn &

    Bacon.

Tapscott, D. (1999). *Growing up digital: The rise of the net generation.* New York:

    McGraw-Hill.

Taylor, C.W. (1988). Various approaches to and definitions of creativity. In R.J.

    Sternberg (Ed.) *The Nature of Creativity: Contemporary Psychological*

    Perspectives, (pp. 99-121). Cambridge, MA: Cambridge University Press.

Toblin, J. (2001). Save the geeks. *Journal of Adolescent & Adult Literacy, 44(6),* 504-

    508.

Torrance, E.P. (2000). (Ed.). *On the edge and keeping on the edge: The University of*

    Georgia annual lectures on creativity. Westport, CT: Ablex Publishing.

Torrance, E.P. (1962). *Guiding creative talent.* Englewood Cliffs, NJ: Prentice-Hall.

Treffinger, D.J., Renzulli, J.S., & Feldhusen, J.F. (1975). Problems in the assessment

    of creative thinking. In W.B. Barbe & J.S. Renzulli (Eds.), *Psychology and*

    Education of the Gifted, (2nd ed., pp. 240-247). New York: Irvington

    Publishers, Inc.

Tyler, K.D. (1998). *The problem with computer literacy training: Are we preparing students for a computer intensive future?* Retrieved May 27, 2006, from http://www.ccs.neu.edu/home/romulus/papers/mywr/report.htm

Tyler-Wood, T., Cereijo, M.V.P., & Holcomb, T. (2001). Technology skills among gifted students: Is there a digital divide? *Journal of Computing in Teacher Education, 18(2)*, 57-60.

Tynan, D. (2007, April 16). The 20 most annoying tech products. *PC World.* Retreived June 28, 2007, from http://www.pcworld.com/article/id,130638/article.html

U.S. Department of Education. (2005). U.S. Department of Education releases National Education Technology Plan. Retrieved June 30, 2007, from http://www.ed.gov/news/pressreleases/2005/01/01072005.html

Van Lieshout, C., & Heymans, P. (2000). *Developing talent across the life span.* East Sussex, UK: Psychology Press.

Van Tassel-Baska, J. (2001). The talent development process: What we know and what we don't know. *Gifted Education International, 16(1),* 20-28.

Walberg, H.J. (1988). Creativity and talent as learning. In R.J. Sternberg (Ed.), *The Nature of Creativity: Contemporary Psychological Perspectives*, (pp. 340-361). Cambridge, MA: Cambridge University Press.

Walker de Felix, J., & Johnson, R.T. (1993). Learning from video games. *Computers in the Schools, 9(2/3),* 119-134.

Wallis, C. (2006, March 27). The multitasking generation. *Time, 167,* 48-55.

West, T.G. (1997). *In the mind's eye: Visual thinkers, gifted people with learning difficulties, computer images, and the ironies of creativity.* Buffalo, NY: Prometheus Books.

White, L. (2000). Underachievement of gifted girls: Causes and solutions. *Gifted Education International, 14(2),* 125-132.

Winner, E. (1996). *Gifted children: Myths and realities.* New York: BasicBooks.

Zimmerman, B. J., & Martinez-Pons, M. (1990). Student differences in self-regulated learning: Relating grade, sex, and giftedness to self-efficacy and strategy use. *Journal of Educational Psychology, 82,* 51-59.

Appendix A

Participant Interview Questions

1. Please tell me about what you are currently doing with computers and technology.

2. What are some of your earliest memory using computers and technology?

3. What sort of external influences would you say strongly affected your interest in computers and technology? Family? Friends?

4. Did you have any computer classes in school? Were you in any gifted classes or programs?

5. Do you think a specific classification in gifted education for 'technology talent' would be helpful? How do you think these students should be identified?

6. Do you prefer to work independently or in groups? Do you enjoy helping others out with computer problems?

7. How would your friends and family describe your personality in general?

8. Do you play computer games often? What types do you like to play?

9. Where do you see the future of technology heading for society?

10. Do you have any last words of advice for young kids out there who are just starting to explore the world of programming and computers and hope to someday be 'techie' professionals?

11. What skills/experiences do you think are necessary for a good 'techie' to have?

Appendix B

Thumbnail Sketches Developed from Pilot Study

Computer Technology Talent: What Is It?
R. Friedman-Nimz and B. O'Brien, 2005
University of Kansas

Thumbnail Sketch A (Programmer)

Sam is a programming student who enjoys working alone with the computer language and spending hours deciphering code and playing with creative programs. He has worked on the school's website since middle school, as well as creating a personal site about his favorite movies and music. He completed all of the available computer courses offered by the school district, but often felt frustrated by the slower pace of the rest of the students in his class. Sam took several classes at the local university in the summer, and he felt that the challenge of college courses was better suited to his skill level. When he was asked to suggest improvements for his high school, he mentioned more advanced classes, more individualized instruction, and more computers available for student use. He identified his technological strengths as good logic, problem solving and programming skills. He socializes well with his peers in the computer club, but when it comes to content learning, he prefers to work independently.

Thumbnail Sketch B (Interfacer)

Jean had always been a computer troubleshooter and enjoyed helping her teachers and peers at school with their computer technology problems. She is less interested in exploring the computer itself (e.g. hardware) and is more into the social interactions resulting from helping people with technology. She expressed that she would become bored with nothing but computer programming all day, and liked to vary her computer interactions. The excitement of unraveling and solving technical issues and improving old technology appears to be what intrinsically motivates her. She had served as an assistant to teachers in middle school and is currently an aide in the high school's computer lab. Jean also extends her skills further into the community by helping neighbors, churches, and family members with their computer questions. When she was asked to suggest improvements for her high school, she suggested students learning information at an earlier age and having more knowledgeable teachers who shared the same computer interests. She mentioned diagnosing, problem solving, and troubleshooting as her technology strengths. She enjoys working on projects in groups, and likes the creative thinking that comes from group work.

223

Appendix C

Pilot Study Interview Questions

The Project

1. In your estimation, how successful was the project?
2. How could we have made it better on the part of the sponsor? On the part of the Computer Programming Club?
3. In what ways do you imagine that you could use what you've learned from working on this project?

Technology Experiences/Interests

4. How did the work on this project build on your previous experiences?
5. Please tell me a little about your history with computing: when you started, encouragement at home, you classmates' part if any, school's part, out of school assistance.
6. Did you take any helpful courses, workshops or other experiences? Please describe. Where they in/out of school?
7. Please describe your technology-related high school experiences, like with video, graphic art, engineering, computer science, tutoring/consulting with teacher and/or students. Any projects in or out of class?
8. What would have made your school experiences closer to ideal (elementary, junior high, senior high)?
9. What are your technology strengths? Programming, graphics, problem solving, web page design?
10. How does technology relate to your interests in and out of school now? Community service? Job?
11. What technology do you have now? What would you have in an ideal world?

Technology Thinking

12. Below is a list of the ways in which people think. Please RATE how each type of thinking is like you (7 = very much like you; 1 = not at all like you).
13. Do you play chess? If so, for how long? Have you played competitively? Earned any titles?
14. Do you play computer games? What kinds? What do you look for in a game?
15. Do you prefer to work independently or in groups?

Appendix D

Thinking Style Rating Scale

Below is a list of the ways in which people think. Please RATE how each type of
thinking is like you (7 = very much like you; 1 = not at all like you).

_____   Numbers

_____   Words

_____   Pictures

_____   Music

_____   Spatial (3-D)

_____   Relationships among people

_____   Movement

_____   Taking ideas apart

_____   Street smarts – practical thinking

_____   Making creative leaps in thinking

Appendix E

Pilot Study Follow-Up Questions

1. Tell me a little bit about what technology interests you pursued in college and if you plan on going into a career in the technology field.

2. What sort of external influences would you say strongly affected your interest in computers and technology? Family? School? Friends?

3. Do you think a specific classification in gifted education for 'technology talent' would be helpful? How do you think these students should be identified?

4. Where do you see the future of technology heading for society in general?

5. Do you have any last words of advice for young kids out there who are just starting to explore the world of programming and computers and hope to someday be 'techie' professionals?

Appendix F

Parent Interview Questions

1. What are some of your earliest memories of your child using computers and technology?

2. How comfortable are you using technology and computers at home?

3. How would you describe your child's personality?

4. How frequently do you or your spouse use technology in your careers?

5. Do you think it would be helpful for your child if a classification in gifted education for 'technology talent' existed?

Appendix G

Teacher Interview Questions

1. What kinds of computer activities are offered in this class? (video, programming, web-design, etc.)

2. What were the criteria for getting into this program? What made these students different?

3. Overall, how knowledgeable and comfortable are you with technology? How confident are you that you can answer technology questions posed by students?

4. What criteria do you use to evaluate students technology skills?

5. How involved are you in the [independent media group]? Do you feel more like an instructor or a team leader?

6. Would you describe most of the students as 'self-taught' or do they learn from teachers and peers?

7. Do they prefer working independently or in groups?

8. How helpful do you think the courses offered at the school are in furthering your students' computer knowledge?

9. Is technology a priority in your school?

10. In the future, what would you like to see offered to the students in relation to technology?

11. Do you think a classification in gifted education for 'technology talent' should exist? If so, how do you think these students should be identified?

12. Where do you see the future of technology heading for society in general?

13. Do you have any last words of advice for young kids out there who are just starting to explore the world of programming and computers and hope to someday to be 'techie' professionals?

Appendix H

Focus Group Interview Questions

1. Tell me about what media/technology projects you have been working on.

2. How did you get interested in taking this media class (or computer class)?

3. Would you have been involved in these media projects if you did not have the technology available at your school?

4. Do you prefer working alone or in groups? Why?

5. Has the teacher helped with his/her knowledge about technology?

6. What do you plan on doing once you graduate? What jobs interest you?

7. Where do you see the future of technology heading for society in general?

8. I am interested in seeing if there is a classification in gifted education for students who specifically have 'technology talent' – do you think that media technology falls under this category?

9. Do you think some people just have a 'knack' for these kind of projects above and beyond their peers?

Appendix I

Computer Technology Talent Study Information Sheet

Researcher: Brenna O'Brien, PhD
  Contact: brennao@gmail.com
  Sponsor: Dr. Reva Friedman-Nimz, PhD
Institution: University of Kansas

Purpose: The electronic era is upon us, and our everyday culture is changing rapidly. "This generation of children who are now in their teens has become so technologically savvy that being passionate about technology is becoming more commonplace," (Cross, 2005). So how can we distinguish the gifted among this new crop of students?

This study focuses on student experiences growing up with computers at home and at school, as well as personality traits that are associated with computer technology talent. Through analysis of interview data, profiles will be developed to help teachers better recognize students who have technological abilities above and beyond their peers.

To Students: I am hoping that I can schedule a time to chat with you either in person or on-line to ask you some questions about your experiences with computers and technology. The interview will mostly focus on how your home and school life influenced your technology interests while growing up, and what kind of activities you are involved in now, both in real life and in cyberspace.

I'd like to take a few minutes at your school to introduce myself and tell you a little more about my study, and give you an opportunity to participate, if you so choose. Thank you for your interest and I look forward to chatting with you soon!

Brenna O'Brien, PhD
Curriculum and Teaching Department
321 Pearson Hall
University of Kansas
Lawrence, KS 66045
816-561-0441
brennao@gmail.com

Appendix J

INFORMED CONSENT STATEMENT
Exploring Computer Technology Talent

INTRODUCTION
The Department of Curriculum and Teaching at the University of Kansas supports the practice of protection for human subjects participating in research. The following information is provided for you to decide whether you wish for you and/or your child to participate in the present study. You may refuse to sign this form and not participate in this study. You should be aware that even if you agree for you and/or your child to participate, you and your child are free to withdraw at any time. If you or your child does withdraw from this study, it will not affect your or your child's relationship with this unit, the services it may provide to you or your child, or the University of Kansas.

PURPOSE OF THE STUDY
As the field of computer technology continues to change and to grow, it is clear that some students demonstrate a specialized ability for learning and understanding the workings of computers, crafting unique applications of computer technology, creating programs, and/or displaying an unusual knack for computer-based problem solving. The purpose of this study is to explore the life experiences and personality characteristics of middle and high school students who exhibit computer technology talent.

PROCEDURES
Interview questions will be asked that focus on the student's history with computers, helpful educational experiences, interests, and self-assessment of pertinent skills. Teachers and parents may also be interviewed, when appropriate, to provide a well-rounded picture of the student. Questions may be followed up through e-mail contact, with the consent of the participants.

RISKS
There are no risks associated with participating in the project. Your name or your child's name will not be associated in any way with the information collected about you or with the research findings from this study. The researcher(s) will use a study number, initials, or a pseudonym instead of your or your child's personal name.

BENEFITS
Anticipated benefits are: improved procedures for school districts to use for identifying students who have computer technology talent, improved access to educational opportunities for students, and information that will assist parents in working with their children and their school to meet students' needs in this area.

INFORMATION TO BE COLLECTED
To perform this study, researchers will collect information about your and your child's experiences with computers. This information will be obtained from: a student, a teacher

familiar with the student's work, and the student's parent(s) or guardian(s). Also, information will be collected from the study activities that are listed in the Procedures section of this consent form.

The information will be used by Brenna O'Brien and Dr. Reva Friedman-Nimz to identify important qualities that comprise computer technology talent and key educational experiences that enhance the development of this talent.

The researchers will not share information about you and/or your child with anyone not specified above unless required by law or unless you give written permission.

Permission granted on this date to use and disclose your information remains in effect indefinitely. By signing this form you give permission for the use and disclosure of your and your child's information for purposes of this study at any time in the future.

REFUSAL TO SIGN CONSENT AND AUTHORIZATION
You are not required to sign this Consent and Authorization form and you may refuse to do so without affecting your or your child's right to any services you or your child are receiving or may receive from the University of Kansas or to participate in any programs or events of the University of Kansas. However, if you refuse to sign, you and your child cannot participate in this study.

CANCELLING THIS CONSENT AND AUTHORIZATION
You may withdraw your consent for you and your child to participate in this study at any time. You also have the right to cancel your permission to use and disclose information collected about you or your child, in writing, at any time, by sending your written request to: Brenna O'Brien c/o Dr. Reva Friedman-Nimz, University of Kansas, Department of Curriculum and Teaching, 321 Pearson Hall, Lawrence, KS 66045. If you cancel permission to use your and/or your child's information, the researchers will stop collecting additional information about you and/or your child. However, the research team may use and disclose information that was gathered before they received your cancellation, as described above.

PARTICIPANT CERTIFICATION:
I have read this Consent and Authorization form. I have had the opportunity to ask, and I have received answers to, any questions I had regarding the study and the use and disclosure of information about me and/or my child for the study. I understand that if I have any additional questions about my rights as a research participant, I may call (785) 864-7429 or write the Human Subjects Committee Lawrence Campus (HSCL), University of Kansas, 2385 Irving Hill Road, Lawrence, Kansas 66045-7563.

INFORMED CONSENT AUTHORIZATION
Exploring Computer Technology Talent

PARTICIPANT CERTIFICATION:
I have read the Consent and Authorization form for this study. I have had the opportunity to ask, and I have received answers to, any questions I had regarding the study and the use and disclosure of information about me and/or my child for the study. I understand that if I have any additional questions about my rights as a research participant, I may call (785) 864-7429 or write the Human Subjects Committee Lawrence Campus (HSCL), University of Kansas, 2385 Irving Hill Road, Lawrence, Kansas 66045-7563, email dhann@ku.edu.

If you have questions about the study, please contact the main researcher, Brenna O'Brien, at 785-864-9724 or brennao@gmail.com.

_____

Print Student's Name

I agree to allow my child to take part in the study. By my signature I affirm that I have received a copy of this Consent and Authorization form to keep.

_____   _____

Parent's Signature   Date

_____

Print Parent's Name

_____

Print Parent's E-mail Address (optional)

By writing my e-mail address I agree to be contacted electronically for any further follow-up questions required for this study.

Researcher Contact Information
Brenna O'Brien, PhD
Curriculum and Teaching Department
321 Pearson Hall
University of Kansas
Lawrence, KS 66045
785-864-9724
brennao@gmail.com

INFORMED ASSENT AUTHORIZATION
Exploring Computer Technology Talent

As the field of computer technology continues to change and to grow, it is clear that some students demonstrate a specialized ability for learning and understanding the workings of computers, crafting unique applications of computer technology, creating programs, and/or displaying an unusual "knack" for computer-based problem solving. The purpose of this study is to explore the life experiences and personality characteristics of middle and high school students who exhibit computer technology talent.

Thank you for agreeing to participate in this study. By signing this form you give permission for the use of interview information. All results will be confidential. You will not be identified in any way.

_____   _____
Student's Signature   Date

_____
 Print Student's Name

By my signature above I acknowledge that I have received a copy of this Consent and Authorization form to keep.

_____
 Print Student's E-mail Address

By writing my e-mail address I agree to be contacted electronically for any further follow-up questions required for this study.

Researcher Contact Information
Brenna O'Brien, PhD
Curriculum and Teaching Department
321 Pearson Hall
University of Kansas
Lawrence, KS 66045
785-864-9724
brennao@gmail.com

www.ingramcontent.com/pod-product-compliance
Lightning Source LLC
LaVergne TN
LVHW062313060326
832902LV00013B/2191